CW00434840

DIGITAL CRAFTS

To Bjorn, Kari and Siri,

Biddy and Joe.

First published in Great Britain 2013
Bloomsbury Publishing Plc
50 Bedford Square
London WC1B 3DP
www.bloomsbury.com

ISBN: 9781408127773

Copyright © Ann Marie Shillito 2013

A CIP catalogue record for this book is available from the British Library

Ann Marie Shillito has asserted her rights under the Copyright, Design and Patents Act, 1988, to be identified as the author of this work.

All rights reserved. No part of this publication may be reproduced in any form or by any means – graphic, electronic or mechanical, including photocopying, recording, taping or information storage and retrieval systems – without the prior permission in writing from the publishers.

Commissioning editor: Susan Kelly
Copy editor: Ellen Grace
Proofreader: Jane Anson
Page design: Evelin Kasikov
Cover design: Eleanor Rose

This book is produced using paper that is made from wood grown in managed, sustainable forests. It is natural, renewable and recyclable. The logging and manufacturing processes conform to the environmental regulations of the country of origin.

Printed and bound in China

Ann Marie Shillito

DIGITAL CRAFTS

INDUSTRIAL TECHNOLOGIES FOR APPLIED ARTISTS AND DESIGNER MAKERS

Acknowledgements

First and foremost, thank you, all you lovely designer makers and artists who have contributed by responding to my original questionnaire and with statements and images; the existence of this book relied upon your generosity. Thank you, Kathie Murphy, for instigating this with Bloomsbury although I did cuss at times.

I have a very special thank you to Malcolm McCullough for writing *Abstracting Craft: the Practiced Digital Hand*, so rich with concepts, information and facts, and with such interesting and relevant reasoned and philosophical threads that my paperback copy, purchased in the US in 1998, is well thumbed and quoted, decorated throughout with my coloured sticky tabs.

Thank you, all my friends, colleagues, and family (especially Bjorn) for your support and tolerance over these past few years as I have really, really needed it. And Davida Forbes, for your quiet guidance and editorial firmness with me and my eclectic style of writing.

Note: Much of the research for this book comes from a questionnaire I sent out in 2010 to designer makers I knew were using digital and industrial technologies in their work. Responses to the questionnaire are used throughout the book.

'SNAPSHOT'.
A snapshot of the work that designer makers' have contributed to this book.

CONTENTS

CHAPTER 5

CHAPTER 6

CHAPTER 7

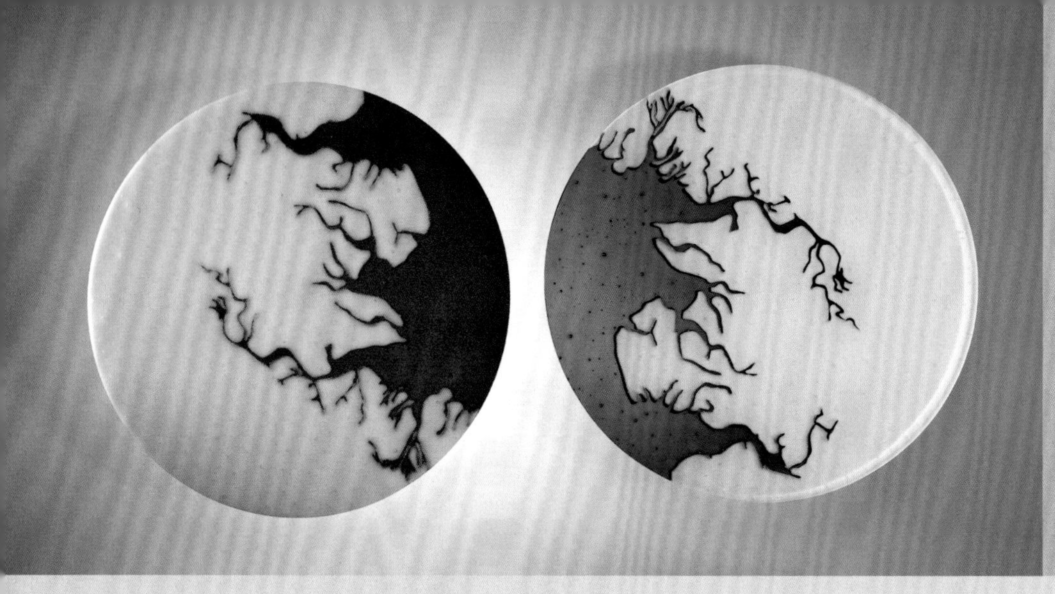

◀ **'MICRO MACRO' BY INGE PANNEELS.** Waterjet cut and fused glass, 2 x 28 cm diameter, 2011. Private collection. Photographer: Kevin Greenfield.

▶ **'VORTEX' BY MICHAEL EDEN.** Made by 3D printing. 2011. Image courtesy of Adrian Sassoon.

It is not craft as 'handicraft' that defines contemporary craftsmanship: it is craft as knowledge that empowers a maker to take charge of technology.
— Peter Dormer, 1997

The most exciting things are the opportunities opening at every level for designer makers to take charge of digital technologies such as computer-aided drawing and designing [CAD], visualisation, 3D modelling and processing applications; at the same time as using industrial technologies such as digital printing, laser/waterjet cutting, computer numerical controlled [CNC] cutting and milling and 3D printing. It is now easier than ever before to explore, exploit and adopt these technologies and apply them to enhance our practice, introduce new ways of working, new business models and, with expansion of social networking, new marketing methods.

Don't be intimidated!

Digital data – as bits and bytes – underlies the pipeline from design to production to marketing because bits and bytes define designs by describing pattern, shape and form. They control what machines do, where they cut and mark, where material is added and where it is not. Most of the objects illustrated in this book have been created by professionals, and for anyone wanting to start on the first rung of the ladder this can be intimidating, and possibly de-motivating. But barriers such as cost and complexity are diminishing. I have selected the illustrated work in this to be inspirational – everyone starts at base level. I want this book to empower, knowing that engagement with and access to digital technologies will continue to improve and that as designer makers we have exceptional knowledge and expertise to take full advantage of all the means available to enhance our practice.

Digital work and craft practices

Digital technologies are there to be used and mastered in the same way as all the other tools and processes in our 'toolkit'. They may seem very different however, as being 'digital' means the processing part is 'hidden', making understanding and controlling the process from concept to end product seem more complicated, unfamiliar ... and definitely not craft.

'Craft' is generally understood to be about controlling the whole process from start to finish, adopting, adapting and improving tools as the need arises. This intimate hands-on approach to making feeds knowledge and information into a referential active loop where experience informs action, which in turn becomes the experience and so on.

'ALBERT CHAIR', DESIGNED BY HECTOR SERRANO. 3D printed in sheets of coloured A4 paper by McorTechnologies Ltd on their Matrix 300 3DPrinter. 2009. Volume: 54 cc. Photographer: Cormac Hanley.

What happens when new and disrupting tools and processes break this experiential loop by removing some empowering knowledge? Being pragmatic as well as inquiring, inquisitive and questioning beings, as designer makers we naturally adapt if these new applications and tools have potential to add value to our practice and will inspire and extend our range of work. This is what is happening with digital technologies such as laser/waterjet cutting and '3D printing' as we collaborate with the service providers. As designer makers we are well aligned and affiliated to digital technologies as we have the 'right approach, skills and mindset' (J. R. Campbell, 2007) to explore the 'close relationship between digital work and craft practice' (Malcolm McCullough, 1996).

To quote J. R. Campbell further, applied artists, craftsmen and designer makers are 'emerging as leaders in exploring the hybrid connections or realms between digital/virtual and hand/real expressions'. Designer makers can straddle the analogue and the digital worlds, and have the resources to combine them so that the results can be far, far greater than the sum of the parts. The work being designed and made using digital technologies is amazing, beautiful and inspiring.

So why have designer makers been taking to digital technologies like ducks to water ever since computing was made available to them?

Malcolm McCullough (1996) suggests there is a close relationship between digital work and craft practice, arguing that hand and brain activities involved in computer use are analogous with making activities involving personal commitment and tacit knowledge. He built his thesis through examining handicraft, design vision, and tool usage as fundamentally human activities, emphasising the importance of this commitment and this knowledge, which is implicit in hand work. I include his argument – that hand and brain activities involved in computer use and in coding are akin to craft practice – because a not insignificant number of designer makers have literally taken control of their designing and making processes by using coding as a tool to explore, to experiment with and to develop different methods to direct their work, and to seek out different aesthetics around their work.

Coding is just another approach to digital designing, another 'tool' that we have for applying our abilities and knowledge as designer makers. Software for creating, designing and modelling is developing rapidly, is easier to use and affordable (some is free!), enabling all of us to join the party. It is liberating to work confidently in the risk-free virtual environment where the constraints of the physical world are no longer limiting factors. In *Crafting Computer Graphics: A Convergence of Traditional and 'New' Media* (2005), Jane Harris argues that as 'the language and working methods of digital practice derive directly from the material practices of e.g. textiles, metal, painting, sculpture and architecture, ... makers in traditional media are ... well placed to develop digital practice.' By being both pragmatic and grounded through the experience of physically working materials, designer makers are able to circumvent digital technologies' seduction to exploit 'virtual world' freedoms. 'A "making" knowledge provides useful constraints in an environment that would seem to have no limitations.' (Jane Harris, 2005). We can apply our knowledge and information, our 'fuzzy logic' right-brain thinking and logical left-brain thinking in practical terms for exploring ideas towards transforming them into tangible expressive artefacts.

■ WHAT IS SO INSPIRING ABOUT DIGITAL TECHNOLOGIES?

To kick start the research for this book I sent out a questionnaire in 2010 to those designer makers I knew who were using 2D and 3D digital and industrial technologies, such as computer-aided drawing and designing, digital printing, laser and waterjet cutting, CNC routing and machining. The following is a sample of their responses to the question 'What is inspiring about these technologies?'[1]

'KINESIS' BY DANIEL WIDRIG. 3D printed, from a collection of laser-sintered, custom-made objects merging notions of architecture, jewellery, and fashion design. The intricate pieces were conceived, generated, and materialised in one seamless digital process. The resulting prototypes were first presented at a catwalk show during the 2012 World Conference on Additive Manufacturing & 3D Printing, hosted by .MGX and Materialise. Private commission. 2012. Photographer: the artist.

Z E-B: *The most inspiring thing about these technologies is that I can now make works that previously I either could not do by hand/conventional tooling or would take so long that they would be physically unviable to make. This alone stimulates a whole breadth of creative opportunities for me adding another layer to my practice. In addition to this there is an amazing amount of fluidity within using digital technologies in terms of inventiveness. Different makers bring different thought processes whilst using the same programs and systems highlighting a massive creative potential.*

TJ: *It enables new creative opportunities, approaches and can help to develop entirely new design ideas and concepts.*

KH: *I use computer-aided design and rapid prototyping alongside my own traditional methods of working as I see the computer as an aid to be able to design and make forms that are difficult or impossible to make using hand processes. I also like the aesthetic forms produced through the structured mesh. Within my work I play with the geometry and facets within the piece.*

CT: *They enable the artist to make different kinds of work at different scales and in multiples. Each piece is an original. It provides all kinds of possibilities and options through the facility to modify and replicate each file.*

DW: *The possibility to generate and materialise form that could not/ hardly be achieved in any other way. The direct and straight-forward production line from file to fabrication with very little manual (= expensive) labour required.*

MA: Pushing boundaries of making work.

GR: *Simulation and testing of designs before the making process. New or different tool sets in a CAD program stimulate new design ideas/solutions. Form can be given to abstract descriptions like mathematical formulae or other sensory inputs. The level of precision that can reproduced at will. That the same digital process can be applied to a variety of materials e.g. laser cutting.*

JS: *New possibilities in production, new aesthetics. Especially the democratic aspect of digital technologies is revolutionary and opens horizons for new relationships between human and object.*

FB: *It is somewhat like standing on the brink of a new world, and being able to contribute.*

LD: *They allow new ways of working that divorce aesthetics from particular making traditions. They allow freedom from 'labelled' practice, Art, Craft, Design, Jewellery, Metalsmith etc.*

1 There is more information on each of these designer makers on pages 149–52.

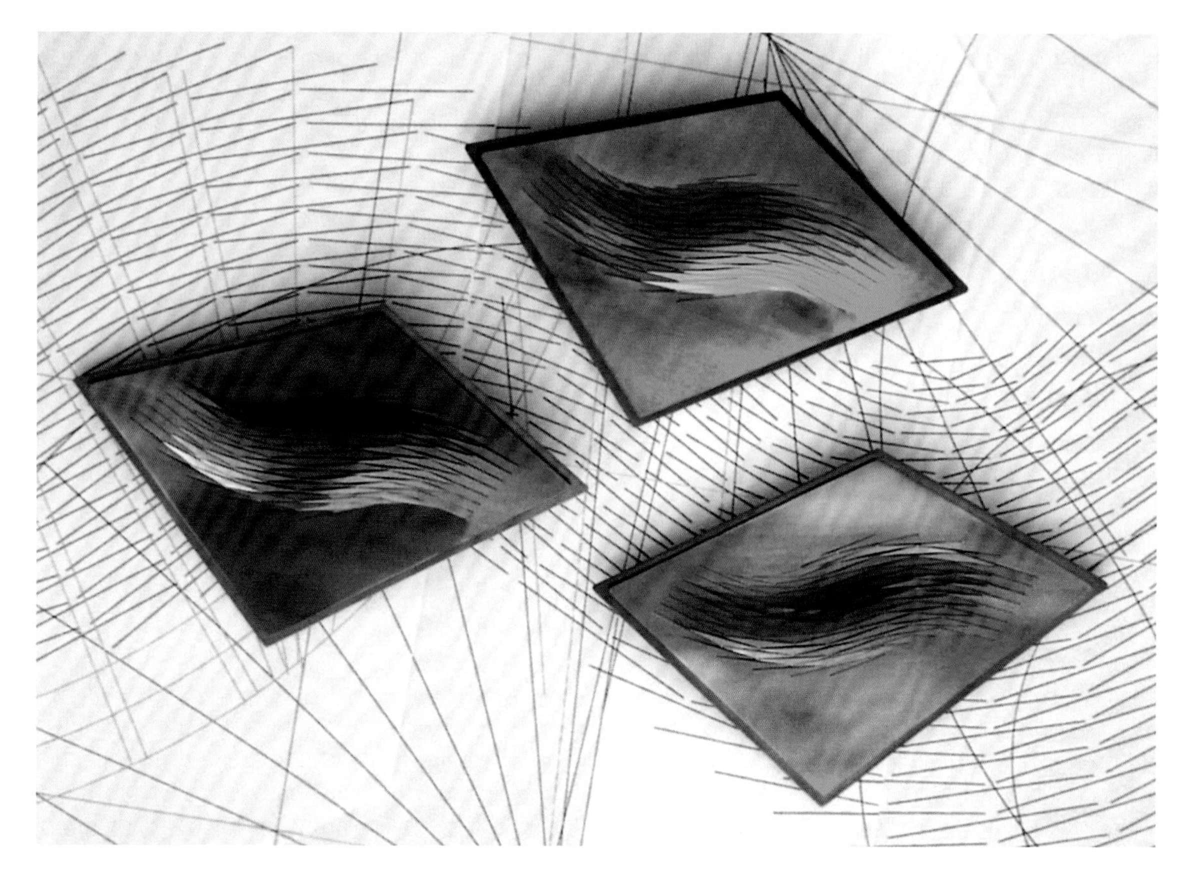

BROOCHES BY ANN MARIE SHILLITO. Designed using CAD. Laser cut and anodised niobium and titanium. Limited edition, 12 x 8 cm, 1992. Photographer: the author.

What this book is about

This is not a 'how to' book, as digital technologies are rapidly changing and advancing. This book is more about providing an understanding of what can be achieved when designer makers embrace digital technologies. I want to augment practice-based knowledge with information that stimulates all levels of thinking and doing to set a virtuous cycle in motion whereby inspiration flows from relevant and easily assimilated information to creating and producing things, which in turn inspires and motivates the next cycle.

I hope to provide enough information about how these pieces came about, for you to decide whether you should commit time, energy and money to digital technologies, and to what level. I aim to assist this process by posing questions about our individual aspiration and capacity to learn a different way of working, to appreciate how digital designing and hands-off processes might complement the satisfaction we get from physically making something. Will the outcome in both the short and the longer term justify the investment, qualitatively and quantitatively? Will we be able to maintain the same sense of autonomy and control over this intangible medium as we can over the physical materials and processes we use, to be able to pile in with confidence, enjoy the process, and achieve a similar level of immersion that we experience at the drawing board and workbench? This is no easy decision, especially as the issues relating to how we choose to make a living are a complex mix of pragmatism and personal subjective choices, and are influenced by things and changes outside our domain.

We have unprecedented access to information through many different channels and the term 'digital technology' covers an ever-growing range of applications. To avoid our being overwhelmed to the point of immobilisation, information needs to be put in context. Furthermore, it should be transferable, reliable, practical and expressed in terms and language that are easily understood and assimilated. This book aims to arm designer makers with motivational and practical information about why and how these technologies can add value to our work, to inspire and stimulate creative thinking and include a wide range of illustrations and examples to show different processes and methods of engagement. I want to encourage a more inclusive encounter with digital technologies and, via healthy disruptive exploitation, establish a means for more designer makers to explore, express and embed new meaning into making.

I have approached researching and structuring this book from my perspective as a designer maker, a contemporary jeweller, who is involved in the development of 3D modelling software designed specifically for applied artists and designer makers.

My own induction into digital technologies began in the late 1980s with a small grant to investigate laser-cutting titanium, a constant material in my studio practice as a jeweller. I laboriously hand worked this difficult but exciting metal: piercing, forging and forming it, all the while searching for tools, processes and technologies for limited production methods for a more economically sustainable business. For laser-cutting titanium and niobium, I discovered that competency in 2D computer-aided design (CAD) is required to generate data for cutting paths.

If it will help others who are thoroughly intimidated by digital technologies, I admit that I struggled to learn to use conventional CAD. I progressed to 3D CAD programs a few years later, in order to explore the potential of 3D printing in materials such as wax, plastics, nylon, and starch. In 2007, together with my research colleague, Xiaoqing Cao, who is a computer software engineer and an amateur artist, we founded Anarkik3D as a specialist software development company. Our 3D modelling product, Cloud9, makes that first step – creating 3D digital models – much simpler, easier – and enjoyable as it uses 'haptics' (virtual 3D touch) as a better way to interact digitally in three dimensions. By exploiting haptic technology we tap into one of our fundamental senses, touch, to make the process of designing in 3D more familiar and natural. In 2011, a personal achievement of mine was designing a wedding ring for my daughter using Cloud9 and having it 3D printed in titanium.

Although I have a high level of knowledge about methods for engaging with digital technologies, it is inevitable that I will not be able cover absolutely everything, not only because of time and space constraints but also due to the furious pace at which digital technology is developing, making it a very 'loose cannon'. My expectation is that both imaginable and unimaginable applications will surface in the future, some useful, others disappearing fast, and the technologies and processes of today will inevitably be subverted or overtaken by others.

I have mitigated this prospect as far as possible by including in each section a cloud of relevant words to use as search tags, as this is more useful than web URLs or QR codes for finding up-to-date information. The glossary of terms on pages 153–4 can also be used for searches. New innovative joined-up ways to find information about how, why, what and where, will appear not just about making and different approaches to creative practice but also about enterprise models to custom-build sustainable practice and business.

Engage and be inspired

Issues around engaging with digital technologies are continually changing, generally for the better. Overall, access is easier and the cost of taking the first steps is plummeting. Usability is improving, too. It is, though, impossible to predict what factors might emerge with significance for designing, making, collaborating, sharing, marketing, selling and managing that justify your investment of time, money and effort. Sections in the following chapters highlight the thoughts and work of designer makers who have taken a digital route. These provide a great starting point – a jumping-in place from which to follow leads and inclinations. This is all that is needed to take the first step.

Be inspired.

2 A craft-minded approach

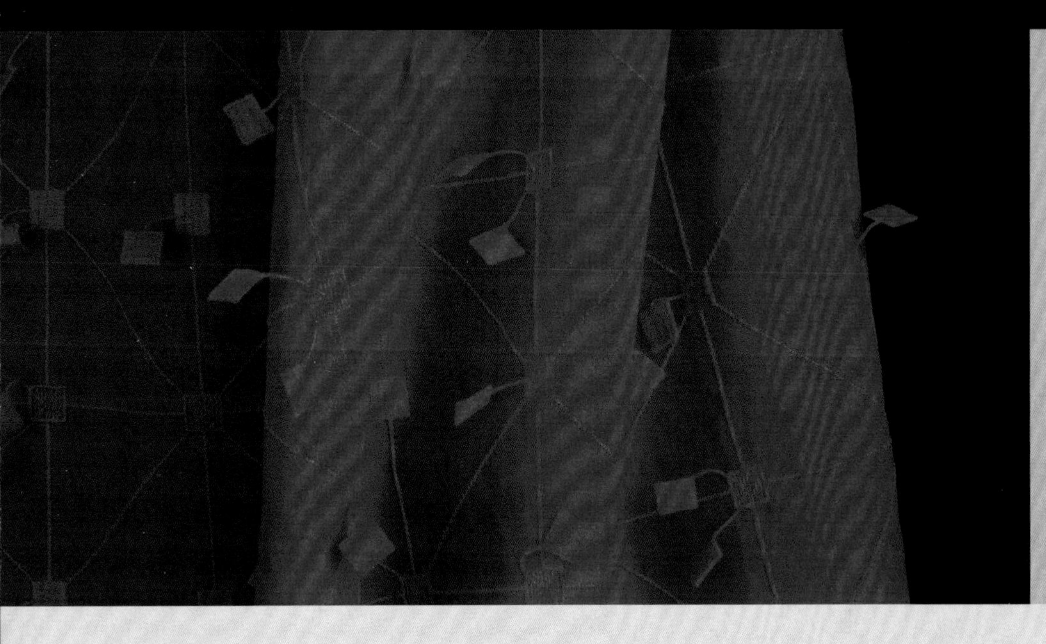

'SWINGING SQUARES' DESIGNED BY REIKO SUDO at Nuno. 100% cotton. Industrial CNC machining. Nuno produce 'small-lot' machine-made textiles working within Japan, calling on the wealth of traditional weaving and dyeing skills and the knowledge and intuition of many brilliant Japanese craftspeople and artisans. Width: 100 cm. Photographer: Sue McNab.

What motivates designer makers to adopt digital technologies?

Most designer makers regard digital technologies as an aid that runs parallel with more traditional methods of working, to be used within our practice as part of our 'toolkit'. Digital technologies enable us to push boundaries, particularly to work through ideas, factors and issues by assisting design, mock-ups and physically making pieces that would be difficult or impossible to envisage otherwise. Moreover, digital technologies enable us to produce new work that was previously impossible, extremely difficult or physically and financially unviable to make by hand with conventional tools and processes.

The amazing work illustrated in this book has been designed using various software programs for 2D and 3D designing and modelling, scanning and digitising, and made using various manufacturing methods such as routing and milling, profile cutting i.e. laser and waterjet, and 2D and 3D printing.

Lionel Dean
INSPIRATION TRIGGER
PRODUCTION PARADIGM ON ITS HEAD
practical advantage
UK CRAFT COUNCIL
EXPLOIT DIGITAL TECHNOLOGIES
J. R. Campbell
MASTERLY FLOW
Richard Sennett
HAPPY ACCIDENTS
CRAFT AS KNOWLEDGE
Gilbert Riedelbauch
BYPASSING TRADITIONAL PRODUCTION METHODS
MAKE TANGIBLE
Beate Gegenwart
PUSH BOUNDARIES
Digital manufacturing
UNVIABLE TO MAKE BY HAND
LEVEL OF COMPETENCY
Michael Polanyi
Designer makers
Farah Bandookwala
RIGHT MIND-SET & SKILLS
mike press
PRAGMATISM AND DISCIPLINE
Justin Marshall
TO ADAPT OR NOT
REAL MATERIALS
FLEXIBILITY AND CONTROL
REAL-WORLD CONSTRAINTS
TACIT KNOWLEDGE
A CRAFT-MINDED APPROACH
DANIEL WIDRIG
MAKING VALUE
Immersive
Peter Musson
SERENDIPITY
ZACHARY EASTWOOD-BLOOM
Kathryn Hinton
POTENTIAL OF VIRTUAL CREATIVITY
PETER DORMER

'LILY091' BROOCH BY JAE-WON YOON combining and contrasting 3D printed structure, designed using CAD and cast in sterling silver, with hand-wrought Mokumegane elements. 2009. Photographer: John McGregor.

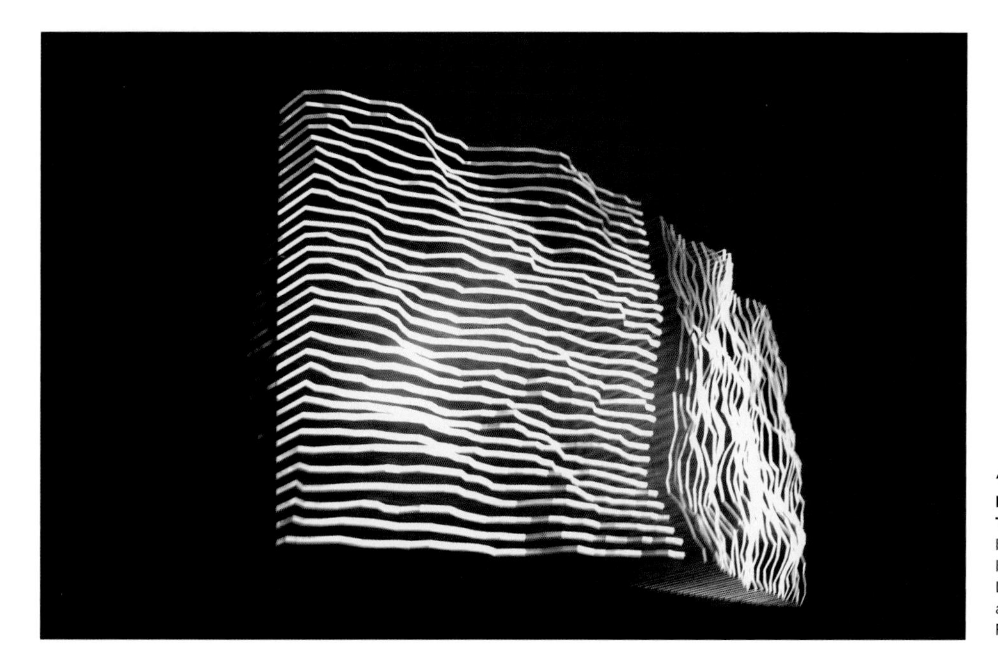

'WE CONTROL THE HORIZONTAL AND THE VERTICAL' BY ZACHARY EASTWOOD-BLOOM. Laser cut Acrylic and amp; Light, 2010. Photographer: the artist

Zachary Eastwood-Bloom is a designer maker who works in London. In his response to the questionnaire I sent out in 2010 he expressed similar views to the other designer makers, stating that:

The most inspiring thing about these technologies is that I can now make works that I previously either could not do by hand/conventional tooling or would take so long that they would be physically unviable to make. This alone stimulates a whole breadth of creative opportunities for me, adding another layer to my practice. In addition to this there is an amazing amount of fluidity within using digital technologies in terms of inventiveness. Different makers bring different thought processes whilst using the same programs and systems highlighting a massive creative potential.

Zac was inspired by Anish Kapoor's sculptures shown at the Royal Academy in 2009. These were made 'using a computer controlled concrete-piping machine. These works showed the interesting dichotomous relationship between digital process and the eccentricities of material properties. They had a really exploratory feel about them.'

Gilbert Riedelbauch is a silversmith in Australia and was hooked the first time he saw the rapid prototype of a minimal surface 'Costa' shape made using SLA (3D printing using a laser beam to set resin in layers). Having a tangible 3D object to directly connect with, demonstrating the importance of visibility, he immediately recognised the immense potential this has for his practice as a silversmith.

These two sides, on the one hand virtual creativity and designing and on the other the industrial technologies, have become bound to each other, although not exclusively, by offering almost unbelievable possibilities that cannot be achieved in any other way. They enable designer makers to digitally generate forms and make them tangible.

Beate Gegenwart is Principal Lecturer and Head of the School of Fine and Applied Arts at Swansea. Her process is drawing on paper, drawing digitally (for instance, in Illustrator), then test cutting on paper with a laser. Of this process she says:

I am aiming for an ever-increasing level of intricacy and complexity, which means many unexpected results and having to very much accept the 'non perfect', every piece is different. At the same time the lasers are able to cut lines and marks finer and smaller than I would ever be able to accomplish in steel by hand. (Momentum Exhibition catalogue, 2011)

'AHX 4' (WHITE) BY GILBERT RIEDELBAUCH. 3D printed in nylon. 227.3 x 203.5 x 181.5 mm. 2001. Photographer: the artist.

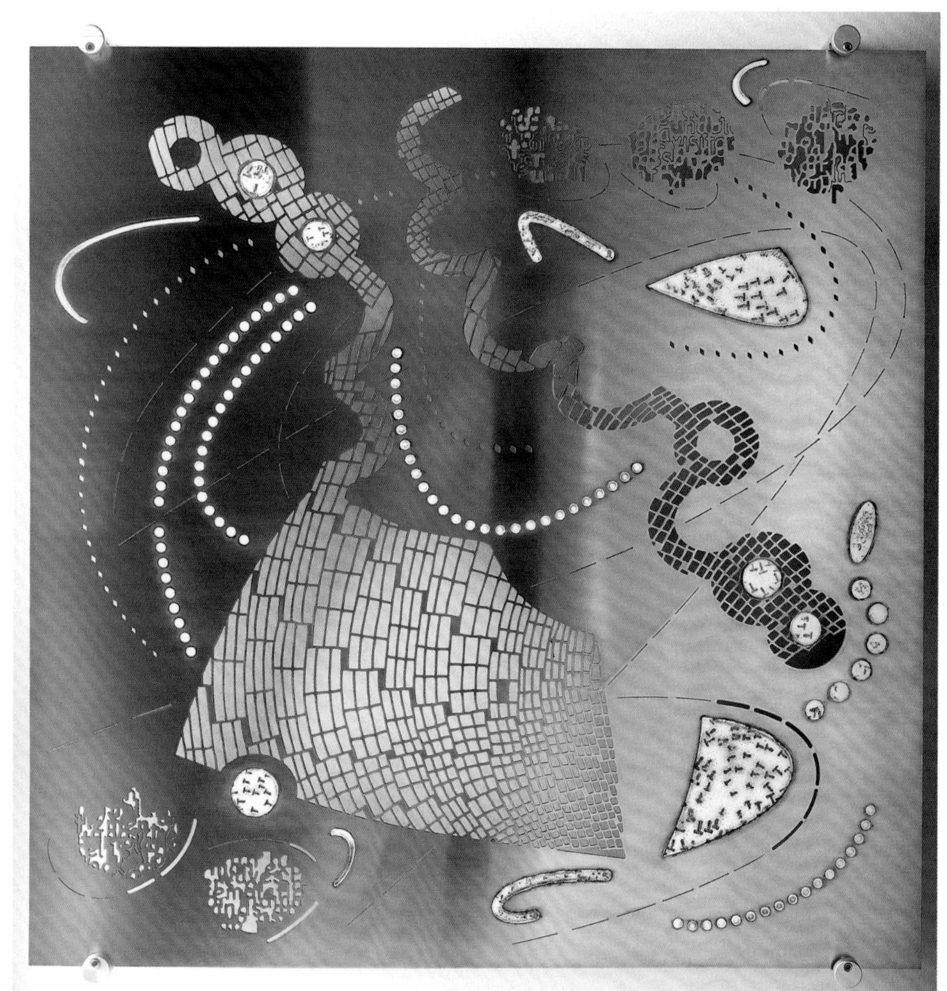

'ARCADE 1' BY BEATE GEGENWART. Vitreous enamel on stainless steel. Photoshop used for masking possibilities; Illustrator used for the beauty of line, drawing and for the vector paths for intricately waterjet cutting the steel. 60 x 60 cm. 2011. Photographer: Aled Hughes.

Craft as knowledge

In their report commissioned by the UK Craft Council, 'Making Value', Schwarz and Yair affirm that increasing numbers of designer makers are accessing digital technologies to explore the potential for themselves. 'In craft discourse, craft is increasingly understood as a distinctive set of knowledges, skills and aptitudes, centred around a process of reflective engagement with the material and *digital worlds*' (my emphasis). (Schwarz and Yair, June 2010)

The software for designing is becoming easier to use, more accessible and user friendly. Some software packages are free, others are expensive and provide high-end features for the serious professional. Most are engineering-based programs and just appearing are products developed specifically for particular sectors and groups, such as JewelCAD and Anarkik3D's haptic sketch/modelling system (Cloud9) for designer makers. In parallel, access to the different 2D to 3D subtractive (laser/waterjet cutting, routing, milling) and additive technologies (3D printing) is more straightforward and in order to attract new customers, service providers for these are introducing a more user-friendly entrée, pushing the costs of manufacture down and the quality of finish up. Available materials span from paper and plastics to titanium and gold, and the end pieces are of a quality sufficient to be final finished pieces, thus bypassing traditional production methods such as stamping and casting and the constraints they impose on design.

However, designer makers are not tied by the constraints that mass production, consumer tastes and price factors inflict on product designers. Lionel Dean, a very experienced industrial and product designer and founder of FutureFactories for the mass individualisation of products, says of digital technologies:

They allow new ways of working that divorce aesthetics from particular making traditions. They allow freedom from 'labelled' practice, Art, Craft, Design, Jewellery, Metalsmith, etc. etc. ...The potential to translate the virtual directly into real-world products still seems magical to me. Digital manufacturing is turning the mass-production paradigm on its head and opening up more possibilities than we can yet imagine. Designs are no longer constrained by process: terms such as concept and prototype become redundant.

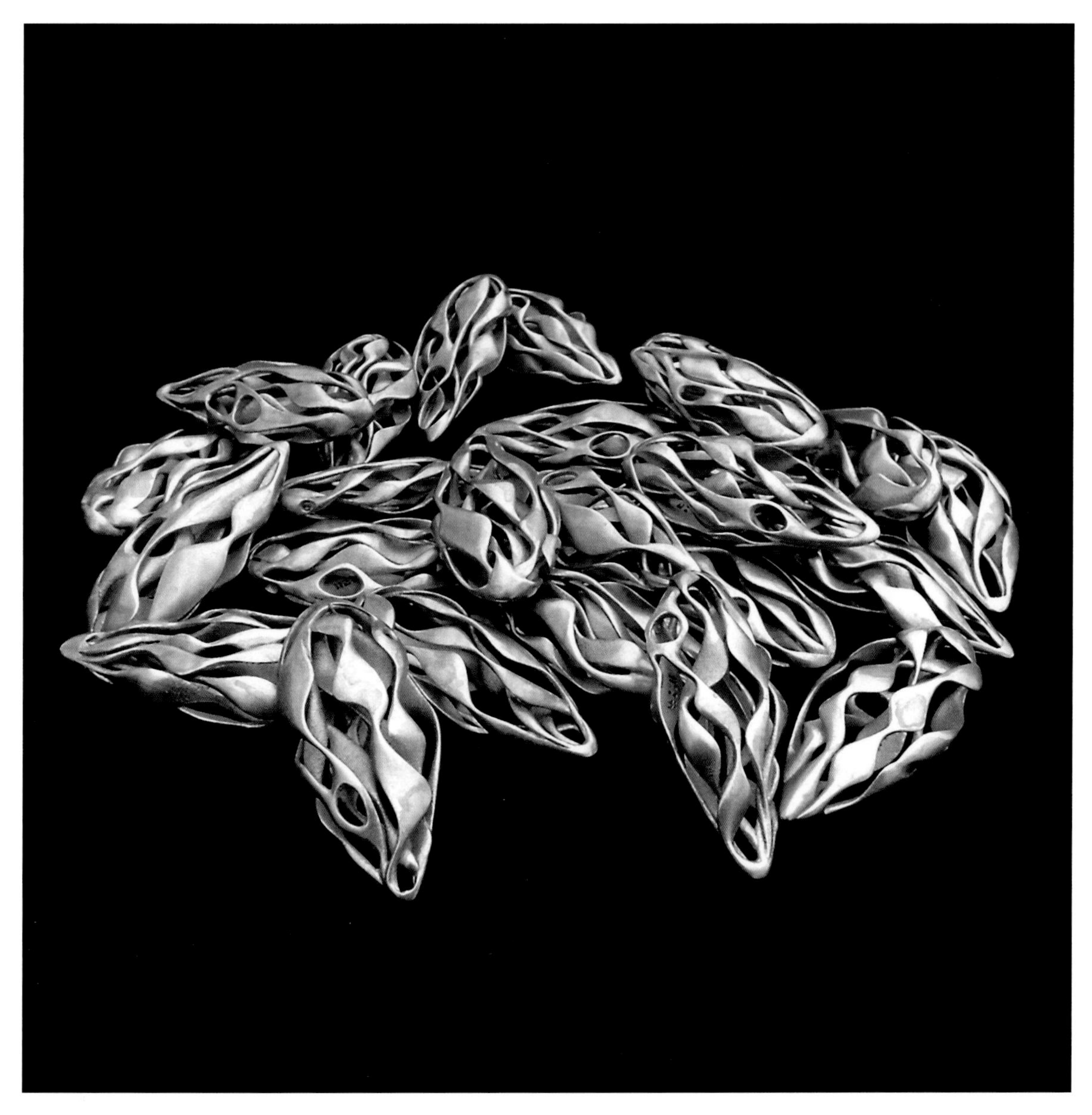

INDIVIDUALISED BATCH OF TITANIUM ICONS BY LIONEL DEAN. 3D printed in titanium. 2008. Photographer: the artist.

Farah Bandookwala, a young New Zealand designer, is particularly excited by the way designers and makers have adapted to digital practices. She mentions Lionel Dean's company, FutureFactories, as a good example of somewhere consumers can take a role in the making process. With its automated, industrial scale production of one-off, individualised artefacts, FutureFactories aims to give consumers ownership of the product by allowing them to select a design or shape configuration from automated iterations. Farah also highlighted service companies such as Shapeways and Ponoko, and says she is fascinated by the way these companies are breaking down barriers in terms of technical skill, and speeding up the development process. She looks forward

'DRAGON' JEWELLERY RANGE BY ANN MARIE SHILLITO. Iterative design process: sketches and hand crafting, piercing and forming prototypes in titanium and steel, transferring know-how into 2D drawing software to develop linked, gradated units. Laser cut in titanium and anodised. 1992. Photographer: the author.

to the day when we might 'desktop manufacture' most objects, rather than mass-produce them in factories. Designer makers have flexibility and control over the choices they make regarding practice and lifestyle. They can choose to work independently, to be self-contained or to work collaboratively; they can take on multiple roles, in teaching, consultancy, projects or research. At times, they may choose to take on other employment, both to mitigate the risky lifestyle that can result from 'pushing boundaries' and to transfer their knowledge into other disciplines and sectors with a high percentage having a 'portfolio practice', as defined in the report by Schwarz and Yair, as work 'beyond the making, exhibition and sale of a craft object'. (Schwarz and Yair, 2010)

'PARASITE' MAGNETIC BROOCHES BY FARAH BANDOOKWALA. 3D printed in polyamide (nylon) and stainless steel, dyed. 2010. Photographer: the artist.

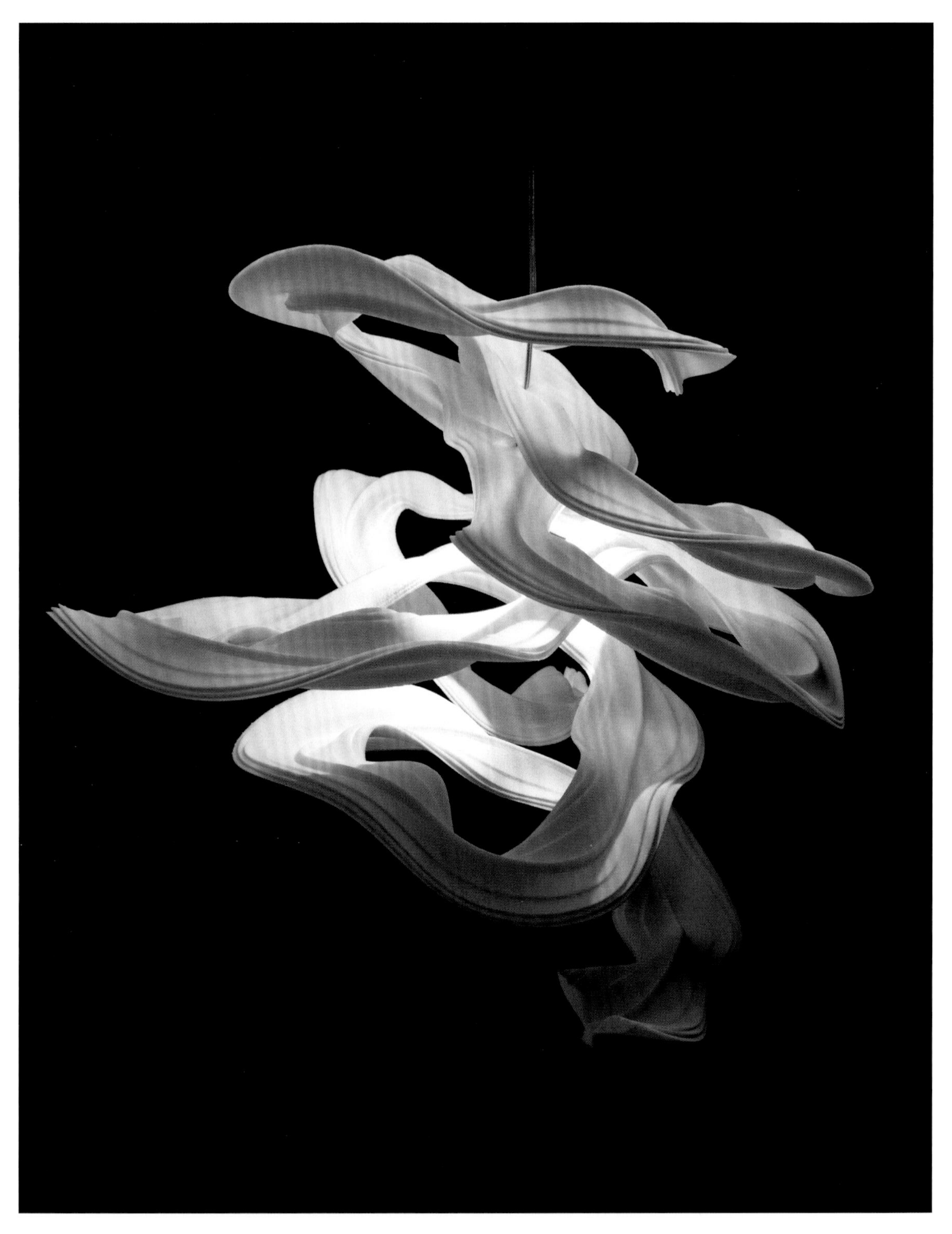

'ATTRACTED TO LIGHT' **BY GEOFF MANN.** Here, the flight of a moth has been captured and made tangible using a combination of cinematic stop-motion techniques, CAD modelling, 3D printing in nylon and traditional hand-craftsmanship. Photographer: Sylvain Deleu. Geoffrey Mann is a Scottish designer artist and the programme director for glass at Edinburgh College of Art. According to Dr Cathy Treadaway, 'his understanding of processes and materials and ability to engage in coherent dialogue with his collaborators is evident in the complexity and success of the work.' (14th January 2011). He is fascinated by the idea of capturing the ephemeral nature of time and motion in works that straddle art, craft and design.

Value and potential of digital technologies to designer makers

As designer makers, we learn to use and adopt new technology where we see its potential for practical advantage and its value to advance and extend our own practice. Researcher Cathy Treadaway recognises how digital technologies 'enable the artist to make different kinds of work at different scales and in multiples. Each piece is an original. It provides all kinds of possibilities and options through the facility to modify and replicate each file.'

The more the computer has become a visual medium, widening the scope for designer makers to break away from the original, formal and constricted methods of digital input/output, the greater its capability and power to disentangle creative practice from the limitations that traditional methods of making impose.

For Daniel Widrig, there are definite advantages:

The possibility to generate and materialise form that could not, or hardly be achieved in any other way, plus the direct and straightforward production line from file to fabrication with very little manual (= expensive) labour required.

From a designer maker's perspective this introduces opportunities for exploring and developing different aesthetics and hybrid practices.

FRUITBOWL BOWL BY DANIEL WIDRIG.
Aluminium. 40 x 40 x 20 cm. 2009. Photographer: the artist.

STERLING SILVER SMALL CAST BOWL 1 BY KATHRYN HINTON. Cast from 3D printed model created by with her 'Haptic Hammer' system. 2011. Photographer: the artist.

Kathryn Hinton uses computer-aided design and rapid prototyping alongside her own traditional methods of working. Within her work she 'plays' with the geometry and facets of the form produced through structured mesh.

I started using digital technology as an alternative method of designing using CAD programs such as Bryce and developing my skills in Rhino 3D. While at the Royal College of Art I was able to utilise the rapid prototyping services and equipment on offer. This enabled me to experiment and test different forms, materials and scales. I was able to use rapid prototyping on a larger scale to integrate with my silversmithing designs.

TESSELLATING HOUSETILE (ONE OF THREE DESIGNS) **BY JUSTIN MARSHALL.** Ram pressed ceramic. 26 x 24 cm (each tile). 2000. Photographer: the artist.

Pragmatism and discipline

Underpinning these positions are the pragmatism, skills and expertise inherent in being a designer maker, a designer craftsman and an applied artist, working directly in materials. Justin Marshall, an independent artist craftsman researching the use of digital technologies by studio potters, writes that:

Craftspeople tend not to use computer technologies to replace existing skills or mimic the nature of pre-existing modes of production. They are used to extend their practices in order to create works that were previously impossible or impracticable to make or even consider. (N.B. the use of these technologies changes the way you think about making as well as your physical practice, it is not just a case of using a technology to functionally provide a new means of doing something.)

MONIKA AUCH: THE MEANING OF DEXTERITY

◀ **'BLACK MUTANT' (DETAIL OF 3D WOVEN FORM) BY MONIKA AUCH.** Material – plastic fishing line, paper yarn. 110 x 45 x 45 cm. 2010. Photographer: Ilse Schrama.

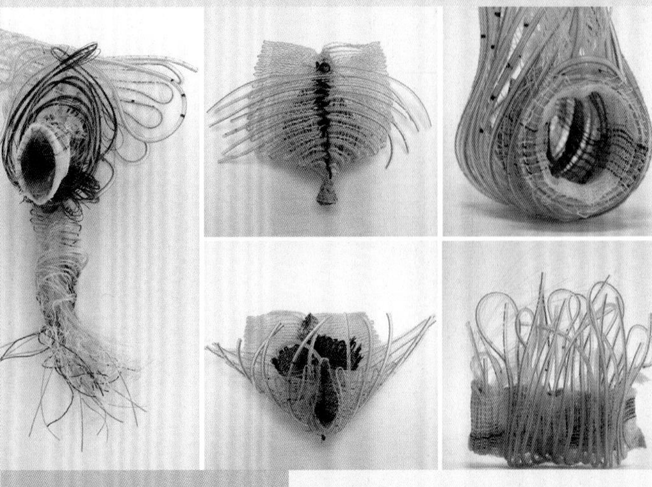

MORPHOLOGY STUDIES: THREE OBJECTS FROM LUDENS SERIES. Left – incompleta. Middle – spina apperta. Right – spina occlusive. Materials – plastic yarn, heat reactive yarn, surgical drains, horsehair. Each about 13 x 5 x 7 cm. 2011. Photographer: Ilse Schrama.

What does dexterity mean in an age which is governed by technological innovations? Are computers, laser machines, 3D-printers and other innovative tools causing a craving for tactility and dexterity as our physical link with the world? In recent years there has been a growing interest in craft-related manual processes of making within the art and design world and the world of technology. Will the increasing digitalisation, in combination with the desire for working with one's hands, put hand and brain at an equal level of appreciation? Or will a decreasing use of our sense of touch trigger the growth of new forms of stimuli transfer within the human body?

In my process of making, hand and brain cooperate with digital tools. Weaving is the oldest binary technique and easily translated into computer programs. I have taken weaving out of its traditional context and used it as a contemporary, autonomous construction technique. A computer-controlled loom is my 'tool' to make sculptures by hand. There is no design of a final product; the work grows by means of a collage of materials, guided by the experience of my hands with tissue structures and with the blueprint of the layered embryonic growth in the back of my mind. Hands are the interface between machine, knowledge and imagination: weaving has an intellectual and a manual side.

In this laboratory of forms I develop cloned series of objects. They are not meant to be true to life copies of existing forms, but additions to what nature had not yet thought of – perfect or not, with construction defects like mutations, limitations and aberrations. All objects in the series are of equal value, there are no failures, just new, unexpected forms. After all, from an aesthetic point of view, irregular shapes can be more fascinating than symmetrical ones. During the process of making there is room for playfulness and serendipity. This working method is directly related to the slow technique. The texture of the work assembles various threads and meanings. The hand is the interface and in this way the weaver is a protagonist of our time, especially in the experience of time itself.

The above extracts are taken from Monika Auch's research and studies, January 2012. Monika studied art and textiles at the Gerrit Rietveld Academy, having first studied and practised medicine. Her fibre sculptures are informed by her medical background and embryologic growth processes.

AWARD WINNING 'PHASE' RING BY LYNNE MACLACHLAN. Dyed 3D printed nylon and (Swarovski) crystals. Lynne MacLachlan is a trained jeweller building virtually in Grasshopper software (Rhino plug-in) using her own coding tool to experiment with graphic structure and form creating shimmering Moiré interference patterns. 2012. Photographer: the artist.

The importance of real craft experience

Designer makers have the right mindset and skills to exploit digital technologies and they also have the right approach to actively explore them. J. R. Campbell maintains that designer makers do have the right abilities:

...to create entirely new product concepts taking advantage of the unique circumstances that the tools can provide, and move beyond using the tools to simply aid in the speed or ease of production... [that] the conceptualisation of new product concepts using digital tools with craft materials, techniques and traditions will increasingly be used to expose not only new markets, but new modes of production and distribution. (J. R. Campbell, 2007)

This defines those inherent attributes that in turn define designer makers and positions them favorably in a post-industrial society where Malcolm McCullough (2204) proposes that close observation and understanding of human experience of the material world and of each other are central issues in designing for the twenty-first century, that it is vital to have knowledge of making in the real world in order to be able to design creatively and effectively with CAD software. .

Richard Sennett describes this grounding as the 'constant interplay between tacit knowledge and self-conscious awareness, the tacit knowledge serving as an anchor, the explicit awareness serving as critique and corrective'. (Sennett, *The Craftsman*, p. 50). Justin Marshall cautions that it is easy to be seduced by CAD's capabilities and forget that it is one element in the production process. Designing something that looks interesting, but is physically unachievable is an easy trap to fall into and has given rise to 'virtual ceramics' that ultimately will never leave the digital realm:

...knowledge and skills are still essential and it could be argued that they are more necessary when working with new technologies. In the virtual environment three-dimensional designs can be visualised with such sophistication that the

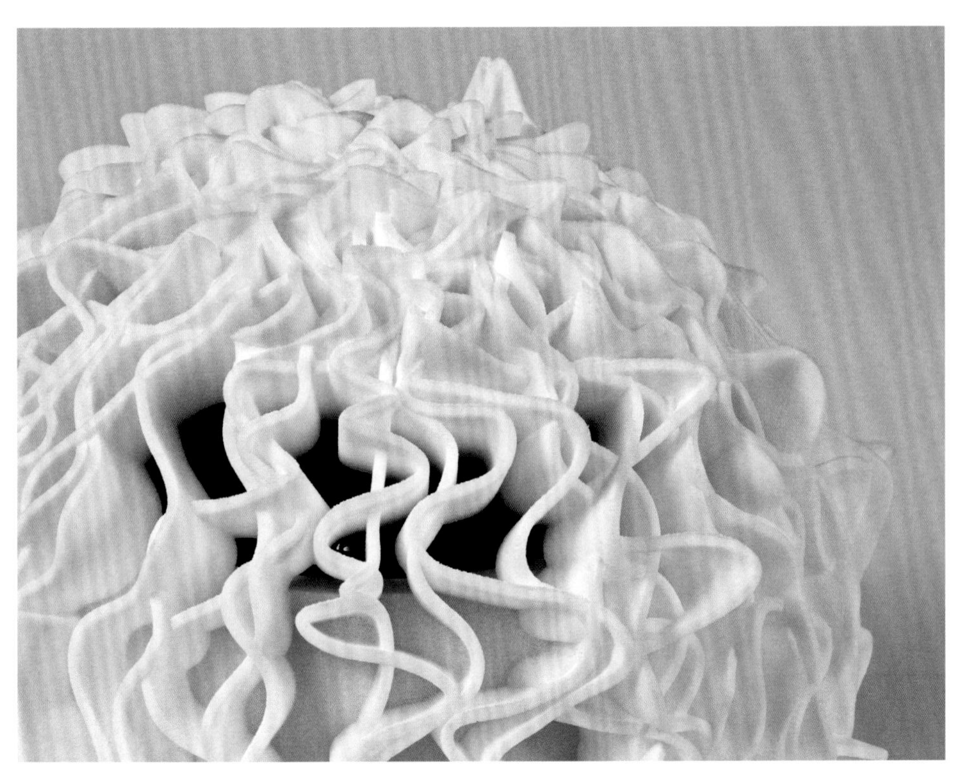

'CLOUDSPEAKER' BY SHAPES IN PLAY / JOHANNA SPATH AND JOHANNES TSOPANIDES. Programming objects using Java-based programming language. 'Processing' as a strong tool offering a new kind of access to design by transferring abstract information, behaviour or movement into shapes and products. 3D Printed. 2008. Photographer: J. Tsopanides.

inherent restrictions of clay can easily be forgotten. The expert ceramist has enough experience of materials and processes to intelligently inform the digital stage of the making process, so reducing the likelihood of wasting time and money designing the unmakable. (Justin Marshall, 2001)

Gilbert Riedelbauch, who is both silversmith and Head of Foundation Studies and the Design Arts Coordinator at The Australian National University's School of Art – and an early adopter of digital technologies – corroborates Justin Marshall's statement, believing that it is particularly useful to be a competent maker in a 'traditional' medium before one takes on digital technologies. The concern of those involved in undergraduate teaching of craft skills, and of those in practice-based research, is that the drive to get students into using computers for concept development and core design work is initiated too early in their course. Eroding the essential space and time required to build up tacit and explicit knowledge through direct experience of real materials can leave many design and applied arts students with superficial and lightweight comprehension of real-world constraints and affordances of materials and processes, that is to say, their properties and qualities that allow a person to perform an action.

Tacit knowledge: a keystone

Tacit knowledge has been put into context and recognised as the base for intuition by philosopher Michael Polanyi and also by Peter Dormer, a contemporary writer on the visual and applied arts, who states:

Craft relies on tacit knowledge. Tacit knowledge is acquired through experience and it is the knowledge that enables you to do things as distinct from talking or writing about them. (Dormer 1997, p. 147). He also writes that: *Disciplined craft is a body of knowledge with a complex variety of values and this knowledge is expanded and its values demonstrated and tested, not through language but through practice.* (Dormer 1997, p. 219)

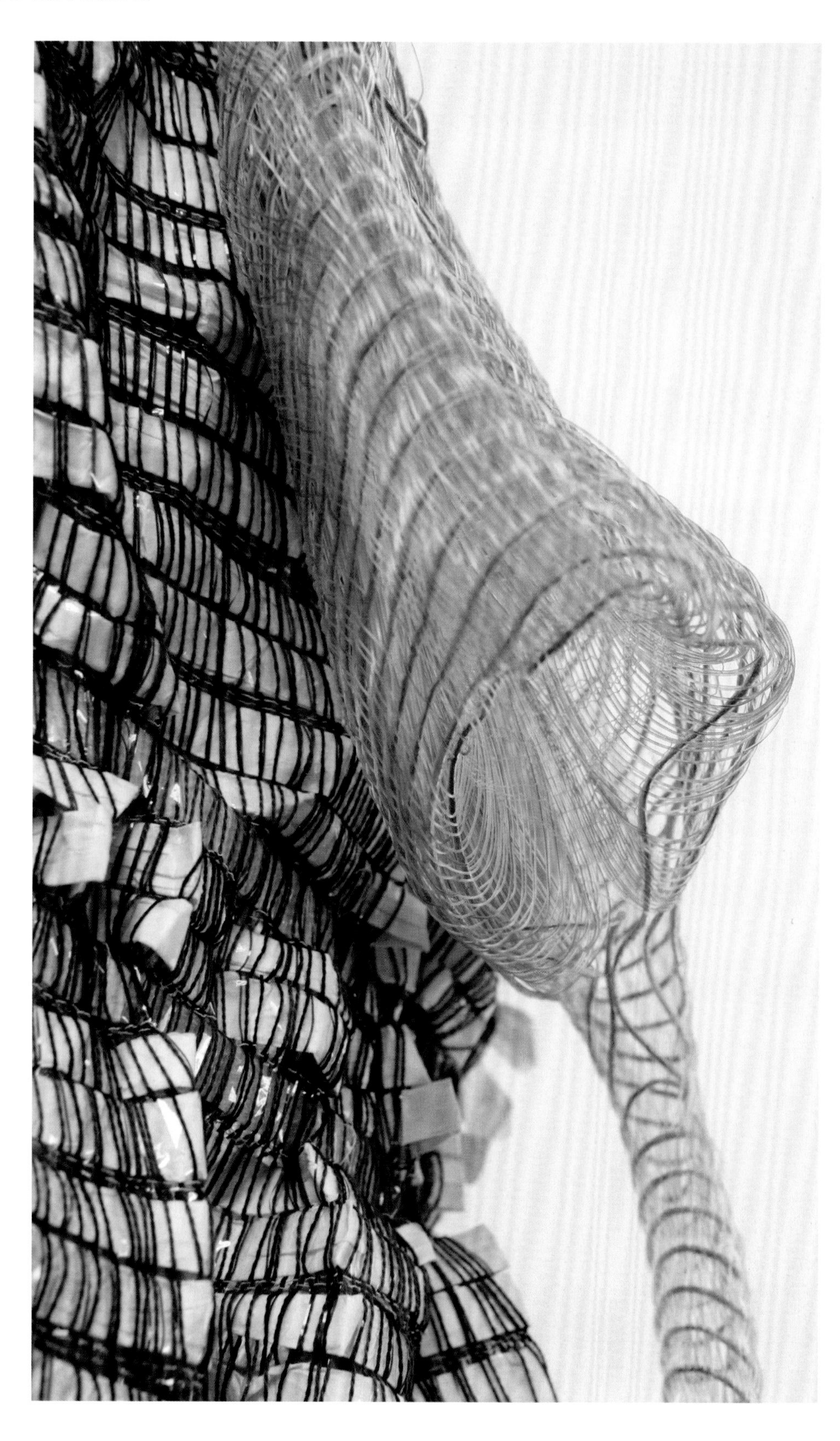

'MUTANT' (DETAIL) BY MONIKA AUCH. Plastic fishing line, paper yarn: this piece taps into 'intricate tacit knowledge, tactile sense and excellent dexterity, only acquired by touching, 'knowing' and valuing the behaviour, properties and emotional meaning of a lot of materials in (weaving) constructions, pushing boundaries by putting them to unusual tasks'. (Monika Auch, October 2010.) 2010 Photographer: Ilse Schrama.

Generally, designer makers find it difficult to write or even talk about their craft with any clarity or coherence, and are not particularly disposed to reflecting on how and why they do what they do. Yet there is a distinct and growing unease amongst them regarding the fast-paced introduction of digital technologies into schools and art colleges: that the spaces and time for physical interaction with materials, tools and process is disappearing, leaving very few opportunities for young people to get their hands 'dirty', to work and make in meaningful and satisfying ways. In 'The Castration of Skill?' (from Tanya Harrod, *Obscure Objects of Desire* [Craft Council, 1997]), Clive Edwards writes:

Understanding or knowing of a process or material cannot simply be discovered by reading, or even doing. Techniques can be learned by 'sitting next to Nelly' but the distinction between 'knowing' and 'knowing how' remains...

'Know-how' comes directly from experience of materials and their physical manipulation with the use of tools. It is difficult to gain this experience any other way than heuristically, through handling and learning, exploration, enquiry, happy accidents and a 'what-if' approach. Where this knowledge can be fully expressed in clear and detailed terms, it is 'explicit' knowledge. Only with practice does this knowledge becomes embedded as tacit knowledge and, as Michael Polanyi explains, 'while tacit knowledge can be possessed by itself, explicit knowledge must rely on being tacitly understood and applied.' (Polanyi 1969, p. 144)

Many people – and I don't mean just designer makers – learn more effectively by *doing*, and they learn most effectively when this is combined with watching a task first done by a 'master'. *Watching* and then *doing* informs at a deep level, to provide both an *implicit* understanding of the totality of the physical act as well as the tacit elements that enable a task to be worked through with mastery and masterly flow.

Is there a dichotomy between 'hands-on' doing and 'digital' doing? Peter Musson describes himself as a craftsman with craftsmanship ideals, and does not believe that skills are ever lost, they just shift and the focus moves on to what is important in the present. He is having a lot of fun playing with the Grasshopper Plugin software, a visual programming language with which he can quickly develop and produce forms for his moulding process saving him from spending hours trying to visualise them or trying to get the modelling right. He recognises this advantage and the fact that he is privileged in having both explicit and tacit knowledge that ensures his digital/virtual model is producible in real-world materials and processes.

'PI' BY PETER MUSSON. Designed using Rhino and Grasshopper (a visual programming language). Made in fine silver formed by positive and negative electro-deposition over a computer-controlled mould. 2010. Photographer: Richard Valencia.

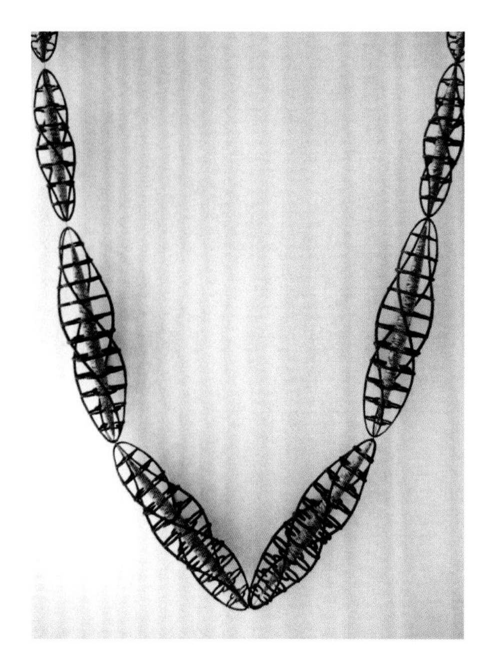

'BUD-4' NECKLACE BY JAE-WON YOON. This necklace used CAD and 3D printing to cast structural forms in sterling silver, from which are suspended mokumegane, metal elements. Traditional and cutting-edge knowledge have been combined to create innovative, new jewellery. 2009. Photographer: John McGregor.

'CIRCLES III' (DETAIL) BY JENNY SMITH, an important creative turning point caused by pressing the 'wrong' button. Laser cut and etched screenprint. 38 x 38 cm. 2009. Photographer: David Gough.

JENNY SMITH: DIGITAL CALLIGRAPHY FRAMED WITHIN THE PARAMETERS OF ZEN PHILOSOPHY[1]

'In a wobbly line we find the truth of pencil'

Roland Barthes

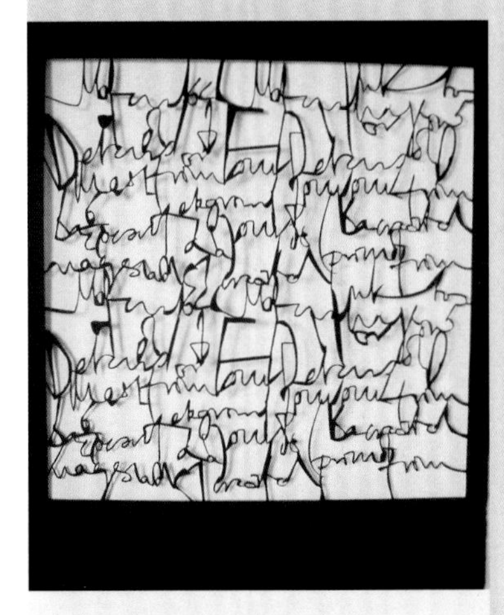

'ZOOM-RESIZE' BY JENNY SMITH. Limited edition drawing, edition of 15, laser-cut somerset paper. 72 x 100 cm. Private collections worldwide. 2007 Photographer: the artist.

Of her first piece, 'Zoom-resize', Edinburgh artist Jenny Smith reflects that 'It took 30 minutes to laser cut, and six months to prepare the files.' She had not realised that laser cutting involved using software and that both have steep technical learning curves.

Jenny is interested in the place that traditional fine art methodologies meet new digital technologies and currently works in the medium of drawing, print, artist books and video. Following a research visit to Japan in 2000 she began investigating the relationship between time, memory and place with particular reference to drawing, framed within the parameters of Zen philosophy. She positioned her meditative, repetitive hand-rendered 'prints', the time-consuming preparation of vector files in Illustrator, and the cutting process, as a creative tool reflecting Zen meditative methods.

By investigating the relationship between time, memory and place with particular reference to drawing, framed within the parameters of Zen philosophy, Jenny places her 'prints' at the point where traditional fine art methodologies meet new digital technologies.

With insight of the key role that chance and accident play within the creative process and working into paper (and not just onto paper), Jenny is interested in the digital equivalent: the interruptions that can occur within controlled processes, celebrated by the Japanese as wabi sabi*. She asks, 'How do I find the wobbly, hand drawn or painted line when using the precise digital medium of laser cutting? And what is the relationship between the concepts that inspire us as artists and the processes we use to investigate them?' Defining drawing as an act of touch when one material meets another, be it tracing a vector file or a laser cutter burning into paper, Jenny is actively receptive, changing direction to respond to the unexpected and the surprising, and subverting the laser cutter's associations with speed, repetition and fixed, anticipated outcome to totally embrace the accidental marks that would not be possible manually.*

(Continued on following page.)

UNTITLED BY JENNY SMITH. Artist's book. Screen printed and laser cut. The surrounding paper and cut-out pieces from *Book of beads* (2009) became a piece of work in itself. 6 x 6 cm. 2010. Photographer: the artist.

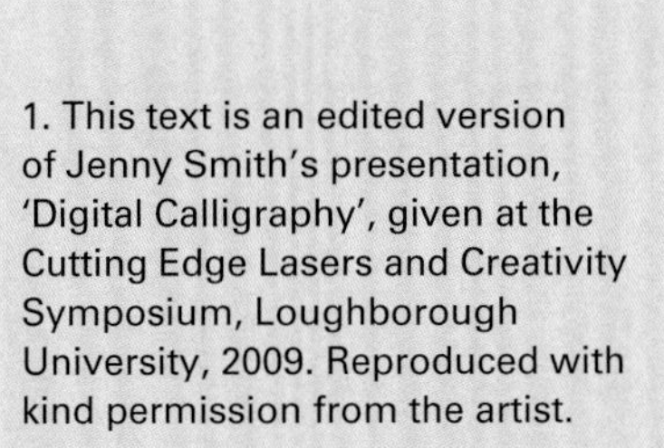

1. This text is an edited version of Jenny Smith's presentation, 'Digital Calligraphy', given at the Cutting Edge Lasers and Creativity Symposium, Loughborough University, 2009. Reproduced with kind permission from the artist.

(Continued from previous page.)

In 'Circles' (2009), paper substitutes paint and digital processes substitute hand painting. By cutting away the negative spaces, the work becomes a form of digital calligraphy, foregrounding the relationship between hand-rendered marks and new technologies. While she was working on this piece Jenny pressed the 'wrong' button in Pathfinder, accidentally creating a file without a border. Only by seeing both hand-cut and laser-cut pieces together did she realise that the 'mistake' was in fact the stronger work, but if someone else had cut the pieces, she would have lost this important creative turning point. She now saves and prints Pathfinder files that don't turn out exactly as anticipated. Similarly, when Illustrator files do not open in Ethos as expected. Jenny uses these 'blips' to learn something new, finding trial and error much more fruitful than any Help menu.

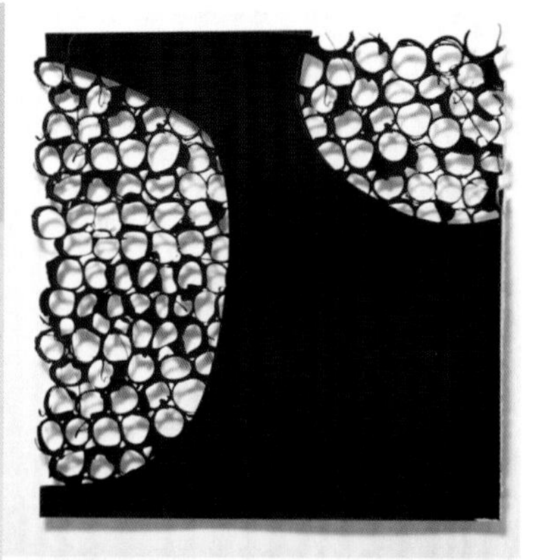

'CIRCLES III' BY JENNY SMITH. Laser-cut and etched screenprint. 38 x 38 cm. 2009. Photographer: David Gough.

In 'Circles II' (2009), an accidental 'kiss' cut line and re-positioned paper that had moved slightly as the laser cutting began created a mis-registered etch line with thin slithers of beige paper visible under the black. This is now one of the signatures of Jenny's work, echoing the first tentative, sketchy lines one makes when drawing that may get smudged, rubbed out or left as a ghost. This was enough to motivate Jenny to buy her own laser cutter!

'CIRCLES XIV' BY JENNY SMITH. Laser-cut and etched screenprint, exploring the subtle differentiations that occur when etching the surface of paper. 38 x 38 cm. 2009. Photographer: David Gough.

Serendipity…

Being able to reflect on exploring and playing initiates new developments and directions and hands-on practice builds new explicit knowledge and creates an environment, which is conducive to serendipitous events and finding valuable or agreeable things not originally sought for.

With hindsight, serendipitous events are often claimed to have 'seeded' an innovation so designer makers are serious about its importance and considering such events to be the trigger for inspiration and aligned with tacit knowledge. Similarly, having the frame of mind to take full advantage of 'happy accidents' relies on the 'know-how' gained from practical experience to project forward and pragmatically recognise the sweet twist it could deliver.

To adapt or not

With computer systems performing faster than we can think and with external pressures to work harder and more efficiently, there is a real danger that effective working practices that nurture the seeds of potential ideas and innovations will have no place within digital practice. Do we work at our own pace on a computer? In *The Craftsman*, Richard Sennett notes that Italian architect Renzo Piano's reason for doing on-site sketching, drawing and model making is because this process of simultaneous thinking and doing facilitates real understanding of the building's relationship in totality with the site's topography. CAD shortcuts this maturing process, as the 'algorithms' in the software complete the drawings and potentially create a closed system divorced from reality. (Richard Sennett, *The Craftsman*, 2008 p. 40)

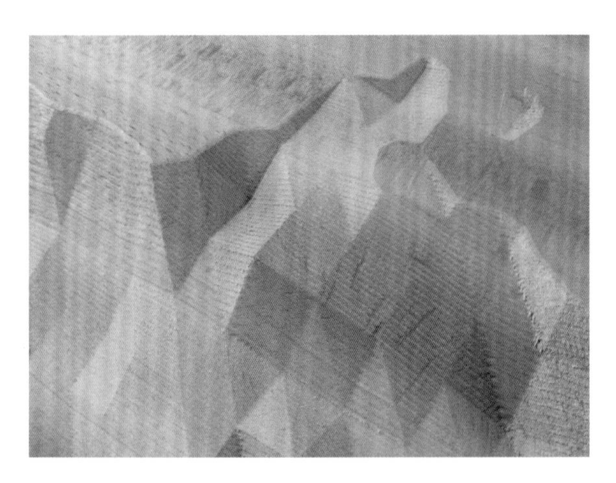

'INFORMATION ATE MY TABLE' (DETAIL) BY ZACHARY EASTWOOD-BLOOM.
CNC milled beech. 2010.
Photographer:
the artist.

New or different toolsets in a CAD program stimulate new design ideas/ solutions. [By using CAD,] form can be given to abstract descriptions, like mathematical formulae or other sensory inputs; levels of precision can be reproduced at will; and the same digital process, e.g. laser cutting, can be applied to a variety of materials.

Gilbert Riedelbauch, silversmith, academic and early user of digital technologies

I have always been an exacting maker and [precision] is one thing digital processes excel at.

Zachary Eastwood-Bloom, artist working in multi-materials

Activity that involves physical things is visceral and can be embedded in the real world and real time. This physicality lends itself to being laid aside and revisited at a later stage, and to 'mulling'. (By 'mulling' I mean occupying the conscious mind at a banal level so that the sub-conscious mind is free to make surprising connections that can lead to quantum leaps and interesting solutions). Computer processing is both invisible and complex, leaving too few discernable handles to mark a good leaving-off point from which to reflect and get a larger perspective. This leads to thoughts and ideas that might not occur within an intensive working session. Often, what seem like time-wasting dead ends but are in fact potentially interesting avenues are only recognised by standing back and seeing the bigger picture.

I am concerned about the 'binary' legacy of early mainframe computers programmed for linear processing, which remains within computer-aided design and graphics packages, and in generative programs, such that the envelope within which digital interaction happens is loaded towards linear methods of thinking. Designer makers do fully embrace this prescriptive way of designing digitally and use it to complement their handwork as its mathematical and engineering bias suits their thinking and practice.

Lighter, more user-friendly design and modelling packages are easily accessible: they are quicker to learn and use, yet sophisticated enough to achieve a range of complex visualisations and forms with less of the undue frustration that is de-motivating. Less structured programmes that allow off the wall practices and exploration, and accommodate different styles of learning and less linear ways of interacting, are still noticeably scarce, but do include technologies like scanning and hybrids such as Freeform and Cloud9 (both haptic systems). As with all skills, once a certain level of competency is reached, ownership happens and rules can be tested and challenged. Writers like Mike Press and J. R. Campbell, who are researching this area are confident that it is designer makers who can and will take on leadership roles in using digital technologies.

3 Virtual for tangible: creating digitally

◀ **'CLOUDSPEAKER' (DETAIL) BY SHAPES IN PLAY/ JOHANNA SPATH AND JOHANNES TSOPANIDES.** Using 'processing' software to access abstract information for design. 3D Printed in SLS by EOS. 2008. Photographer: Johannes Tsopanides

PART 1

Generating, exploring and exploiting digital data

Unlike product designers, designer makers are not tied to the need to digitise their designs. They can choose to use computers where they offer advantages over traditional tools, such as the ability to experiment in a risk-free environment, the freedom to push boundaries, and access to industrial technologies such as 2D and 3D printing, cutting, milling and machining.

Work is created digitally and managed within the context of practice, personal ethos and motivations, and as there as many different combinations of methods and 'making' systems as there are designer makers, I hope this book's structure does justice to their diverse approaches.

JOHANNA SPATH

GETTING DIGITAL DATA

STITCH MY BRAIN

WIMP interface

WORKING 3 DIMENSIONALLY

spatial information

Diverse approaches

PLAYING, CREATING

context of practice

Bill English INTERACTION Immersive

Monika Auch

FAMILIAR WAYS OF INTERACTING

generating NATURAL ABILITY

THE INTELLIGENCE OF THE HAND

digital forms PROPRIOCEPTIVE

INTUITIVE 3D designing

JOHANNES TSOPANIDES

VIRTUAL for TANGIBLE

INTERFACES 3D PRINTING

Alison Bell

PROCESSING

RISK FREE ENVIRONMENT

USABILITY KEY DRIVER Tangible

INDUSTRIAL TECHNOLOGIES

Creating and designing

John Hirschtick, SolidWorks

COMPUTER AIDED MACHINING

sensual DOUG ENGELBART

MACHINING

TOUCH SENSITIVE

LYNNE MACLACHLAN

experiences LASER

WATERJET CUTTING

HAPTICS 2D designing

gobbledygook

Lionel Dean

RICCARDO BOVO

Fred Baier

digitising pens

GESTURE CAPTURING

GUI

motion sensing

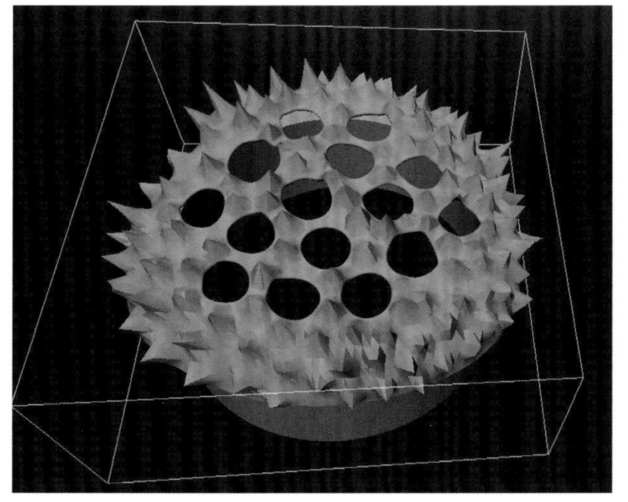

'BRISTLE' BY FARAH BANDOOKWALA, an interactive 3D printed sculpture, fitted with electronics (Arduino), moving parts and sensors, and programmed to be activated by movement. 3D modelled/designed using Cloud9 and Rhino (CAD) software. 3D printed in polyamide and painted. Made as part of her Jerwood Makers Award 2011 Exhibition. Screen capture from Cloud9. Anarkik3D. Photographer: Tomas Rydin.

ALISON BELL: THOUGHTS ON USING DIGITAL IMAGING IN TEXTILE ART

▲ **'BERNERAY 3'** BY ALISON BELL. Digital file: Photoshop. 'Keeping work relatively simple, as I try to find my own language within this new world of perfection'. 30 x 18 cm. 2002. Photographer: the artist.

▶ **'ROCK'** BY ALISON BELL. Digital file: Photoshop. 80 x 20 cm. 2003. Photographer: the artist.

Alison Bell is a textile artist who uses traditional techniques and first approached digital technology out of sheer curiosity. In 2002, she decided to buy a drawing tablet as she could not use a mouse freely enough to draw digitally, even in monochrome. She experimented with various forms of printing images onto fabric, from heat transfer to putting various weights of silk through her HP printer, and although the results were always surprising, the print quality was amateurish. However, she enjoyed this phase, working with the results whatever happened, adding pigments, watercolour and stitch to enhance the printed image.

In Tilburg in the Netherlands, she visited Wilma Kuil, who used photographic imagery that she digitally embroidered. Alison was truly inspired yet, when she returned home was frustrated at her own disappointing printing results. She began working with Glasgow School of Art's Centre for Advanced Textiles (CAT) and a first work was a scan of hand knitting, enlarged and printed onto silk satin. She remembers the hairs on the back of her neck prickling as she watched it emerge from the printer: perfection, magnified, five inches of knitting to 72 inches of printed silk. She was instantly seduced by the sheer beauty and result of the process that hung in her studio for several months, not knowing what or how to take it forward and not touching it for fear of spoiling it. Alison had hit a creative wall.

She was also discovering digital photography and developing simple skills in Photoshop,

'BEGINNINGS' BY ALISON BELL. Digital file: Photoshop. Begin to play with surface textures, exploring layers, revealing depths and subtlety of tone. 110 x 105 cm. 2004. Photographer: the artist.

manipulating images of the shoreline of the Scottish island Arran, where she was living at the time. Some interesting results combined precise photographic images with new layers of digital process and subsequent overpainting in pigments. These digitally collaged pieces, printed by CAD onto different silks, explore surface qualities. The occasional flaws in the printing were exciting, offering ways to make the work her own again – something about the perfection of digital printing had made creative intervention difficult for Alison, seemingly a common issue among textile artists.

Alison kept Photoshop intentionally simple; it is a tool, like a brush, overlaying line, texture, tone and digital painting. She experimented with specially treated and untreated silk, documenting all explorations in a large sketchbook, pushing boundaries, using any happy accidents and not being thrown by mistakes in the printing, but seeing everything as a learning curve.

The idea of illusion was intriguing; digital stitch and real stitch, playing around with low relief, adding small textural 3D elements to the images; her work was slowly evolving. Also, reflection was becoming increasingly important as she built up layers of thoughts onto the silk itself. Analysing her creative process enriched learning, pushing her in unexpected directions – to textile installations of 3D forms in silk, which rotate slowly in their own space.

Working digitally is more about choice than chance. This can be quite baffling at first and is almost crippling to creativity. By

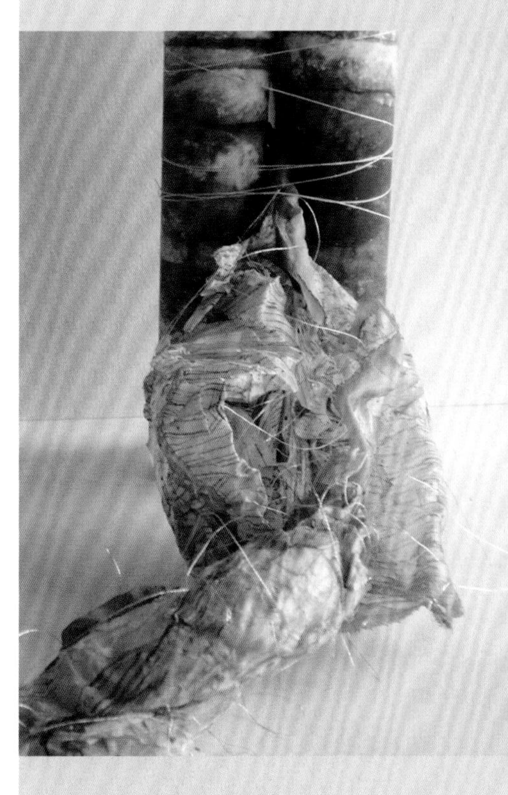

'CYLINDER WITH SILK' BY ALISON BELL. Digitally printed paper, painted, manipulated silk habotai, gold and linen thread: using digital not as an end in itself but a tool to help see into sources of inspiration to draw out and visualise hidden elements. 20 x 30 x 12 cm. 2010. Photographer: the artist.

'SEAWEED & SILK' (DETAIL) FROM 'SEASILKS' INSTALLATION BY ALISON BELL. Created using digital technology as well as more traditional techniques: imagery manipulated in Photoshop, digitally printed on silk satin and overpainted with pigments, then formed by hand. (The seaweed is real!) 18 x 10 x 8 cm. 2008. Photographer: Andrew Stark.

refusing to accept the seductive perfection of digital technology, Alison keeps ownership of her work by embracing risk, absorbing the image back into the creative process. Reflecting about process and crossing boundaries of preconception has enabled her to capture the energy of a place and through flexibility of process, combining techniques and technologies, she can backtrack quickly, manipulating the work until it conveys her meaning – her hand has the final say, and this is a very comforting thought.

This text is an edited version of Alison Bell's statement, sent to the author for this book.

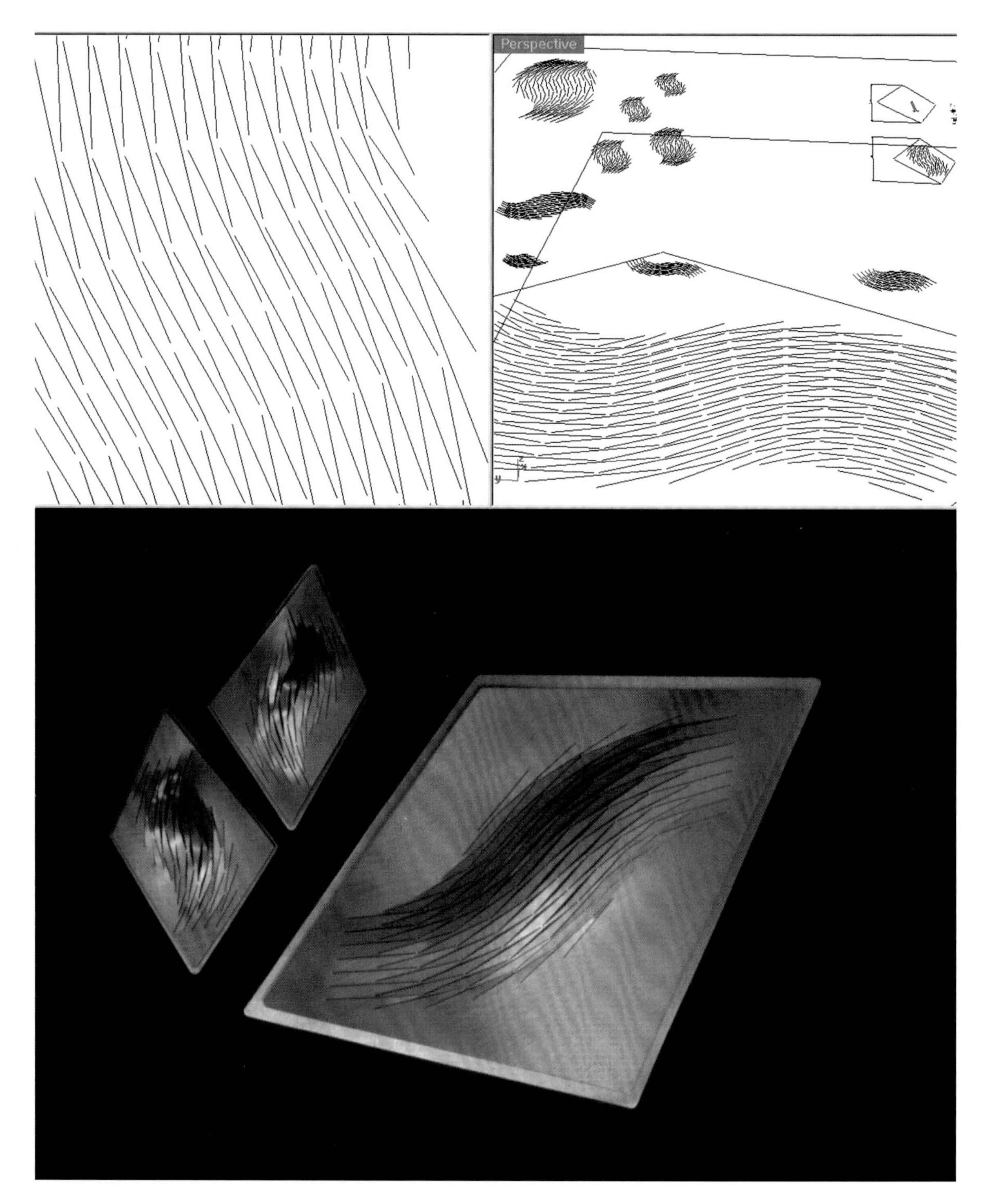

'SLOTTED' BROOCH AND EARRINGS BY ANN MARIE SHILLITO. The top image is a screen capture of the digital designs in CAD (Rhino). On the left is the top view of the lines to be laser cut, on the right is the perspective window. The CAD data for these lines and the outer profile of the brooch were saved in DXF format (data transfer file) for processing for laser cutting. The main image shows laser cut niobium for the front of a brooch and earrings (formed by expanding cuts). Titanium was used for the frame and back (with an integral pin and catch), 12 x 8 cm, 1991/1992. Photographer: the author.

Aquiring data

You need digital data to access industrial technologies, whether 2D or 3D, and although graphics and CAD software are the main tools for generating digital forms and images, other methods include scanning, scripting, mathematical processing or combinations of these.

Digital data controls laser cutters, 3D printers and machines. When designing to produce tangible artefacts it helps to know a bit about different systems, methods, formats, transferring data, how data is processed, how makers innovate by manipulating data at points in this pipeline and 2D methods (printing, machining, laser and waterjet cutting) and for 3D methods (printing, building and computer-aided machining).

Designing with 2D and 3D systems

Fortunately for computer usability the humble mouse was invented. Doug Engelbart sketched the concept in the 1960s, Bill English prototyped it and developers adapted the GUI (Graphical User Interface) to function with the mouse's cursor. For drawing and modelling this was a major leap from typing instructions on a keyboard – as Fred Baier said in the 1980s when designing his iconic 'Prism Chair' (V&A Museum Collection, London) on a computer: 'modelling involved typing gobbledygook for ages until an image finally appeared on screen.' ('Lumps of Geometry', http://www.vam.ac.uk/b/blog/sketch-product/lumps-geometry, Glenn Adamson, 2010)

The mouse has made designing and visualising on computers more inclusive and we have readily adapted to mentally translating horizontal 2D mouse motion to the 2D vertical plane of the computer screen. As digital engagement becomes more closely aligned to real-world type interactions using touch pads and screens and gestures, the mouse cannot compete. For three-dimensional work we certainly need to replace the ubiquitous 2D mouse with more coherent methods of interaction as many of us (including myself) struggle to translate horizontal 2D gestures into the complexity that is 3D.

New applications and concepts for playing, working, communicating and interacting to fit different usability needs are driven by sectors such as digital gaming and the military. Think tracker balls, 'space mouse', tablets and stylus, voice control, digital touch, interactive screens, hand-held 'wands', tiny wireless flecks that can be embedded (in fabrics, under the skin, etc), or capturing hand and body gesturing, tracking eye movement, tapping into brainwaves using electrodes, the list goes on and the key driver is usability.

Developments in technology for creating digitally

John Hirschtick, co-founder of SolidWorks and Group Executive, made four predictions in 2009, one of which was that touch and 'motion user interfaces' using hand movement and gestures to manipulate, model and view forms and data would lead to more natural ways of working with important implications for using CAD. By going back to and exploiting this instinctive, natural ability we tap into intuitive and tacit interactions, more aligned to innate hand-eye coordination where our eyes reconnoitre available spatial information for our hands, pinpointing the target before our hands even start to move. Gesture-based interactions use our proprioceptive sense (see glossary), and with haptics (virtual touch) whether 2D or 3D, fingertips or stylus with force feedback, the whole interactive experience becomes more immersive as disrupting cognitive challenges are eliminated. We have less adapting to do. We can concentrate on learning and creation.

As hardware miniaturises and technology moves towards tapping directly into our very thoughts, will designer makers be involved with the development of applications? Will designer makers be valued for their fundamental and deeply embedded awareness of physical interactiveness gained through the sensual experiences of touch, sound, stereo vision, proprioception and kinaesthetic?

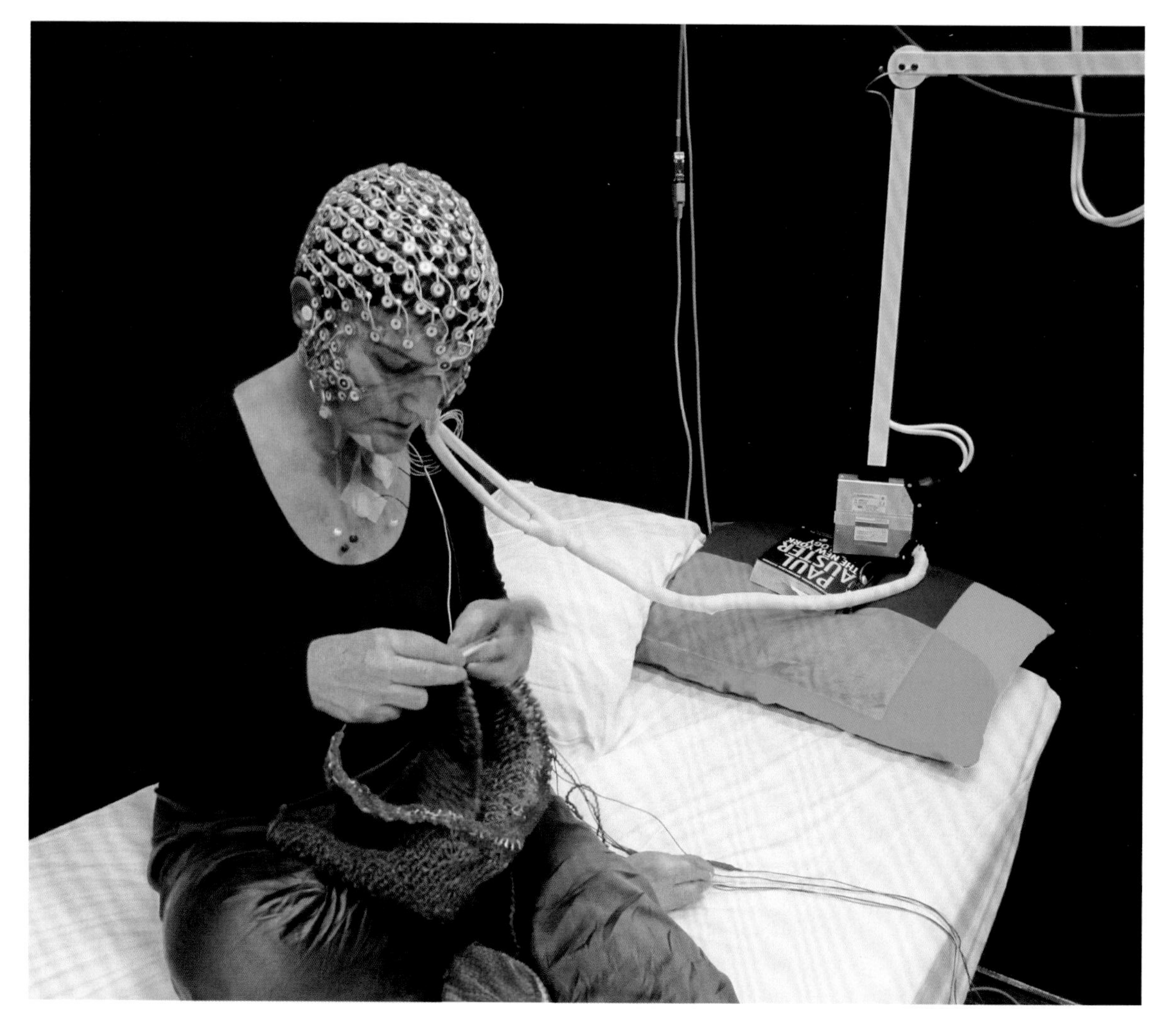

An image of Monika Auch all wired up and knitting for her own research project, STITCH MY BRAIN. Photographer: unknown

To further her own research into how digital technologies influence human creativity, maybe shaping new neuropathways, Monika Auch's project – STITCH MY BRAIN – proposes to define 'the intelligence of the hand' via MRI scans and EEG registrations, mapping the sensitivity of fingertips and dexterity through scientific data.

Designer makers such as Lionel Dean of FutureFactories, Riccardo Bovo, and Johanna Spath and Johannes Tsopanides of SHAPES iN PLAY are involved in new ways to combine apps and hardware to facilitate 'personalisation' by transform and manipulating data, using algorithms, scanning, fingerprints, vibrations from voice, 'blowing', etc. to create dynamic personal forms that can be made tangible.

Meanwhile, the standard computer 'WIMP' interface (window, icon, mouse, pointer), and keyboard can be enhanced with peripherals such as 3D mouses and digitising pens, touch-sensitive pads and screens and input devices with motion sensing and force feedback transposed from video games to 2D and 3D designing.

◀ **'SOUNDPLOTTER' BY JOHANNA SPATH AND JOHANNES TSOPANIDES.** Java-based programming language 'processing' used to transfer captured abstract information (behaviour, movement, sound) into shapes and products. 3D printed in SLS by 3DGiesserei Blöcher. 2008. Photographer: Johannes Tsopanides.

▼ **HOLY GHOST ITERATIONS 1–3 BY LIONEL DEAN,** creating new back and arms for iconic Stark/Kartell Louis Ghost Chair using 3D printed buttons with spring-like links to conform to body shape. The schematic diagram of Holy Ghost Generative script created in Virtools: building block strategy/morphing strategy combined to manipulate/randomise designs by scripting. 2006. Photographer: John Britton.

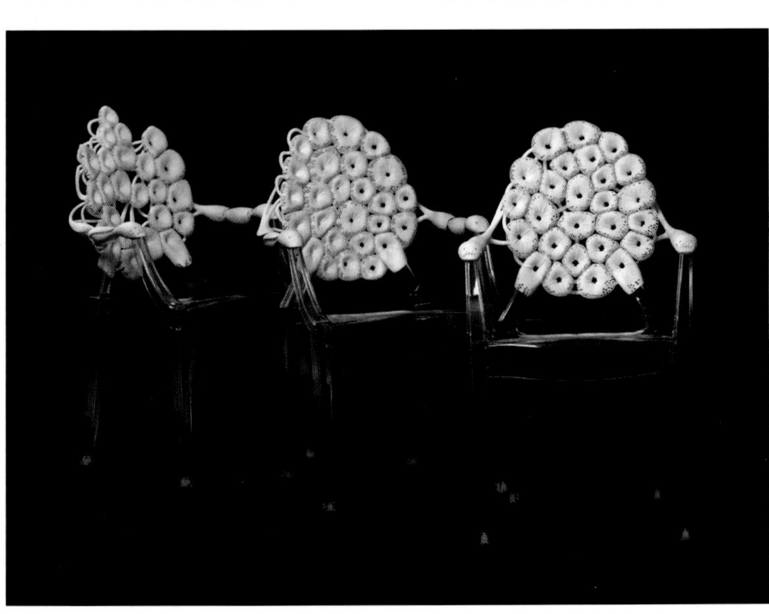

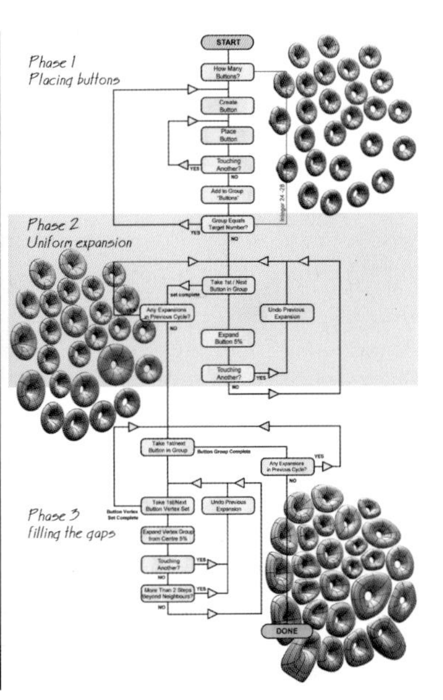

■ LYNNE MACLACHLAN: DIGITAL FABRICATION AND GENERATIVE DESIGN

FLUTTER BANGLE, silver and cold enamel, 10 x 10 cm, private collection. 2008. Photographer (for all images on this and opposite page): the artist.

Lynne MacLachlan's early art school education was in craft-based metalworking where the tools and techniques themselves stimulated what she designed and made. Later a degree in engineering and work experience with graphic designers contributed to making the computer part of her jewellery design practice and from her undergraduate dissertation she came to understand the computer as a digital 'toolbox'. This was probably galvanised by Malcolm McCullough's book Abstracting Craft.

Recognising the potential for a simple computer program to streamline the repetition and repeating patterns of her degree work, Lynne discovered generative art and design on the blog Generator X (http://www.generatorx.no/). While the aesthetics appealed to her, more importantly the program made the generating process visual. Lynne's final work for her master's degree at the Royal College of Art drew on the formation of natural morphologies as a way to synthesise jewellery and objects digitally in parallel with analogue processes.

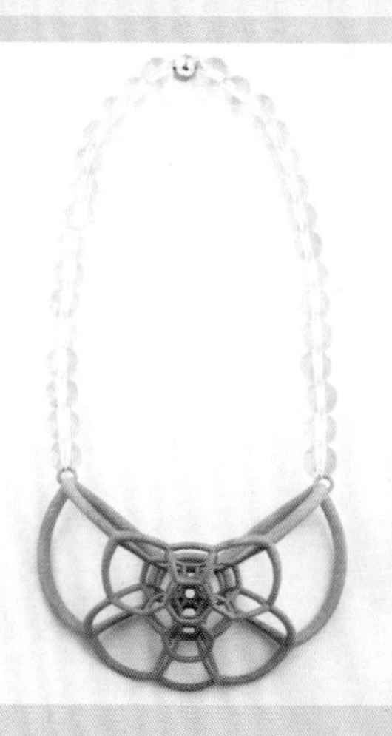

◀ **BUBBLE NECKPIECE,** Nylon/Aluminium mix. 3DPrinted (SLS), rock crystal, silver, 15 x 25 cm, private collection. 2011.

▶ **BUBBLE PENDANT,** silver cast from wax 3DPrinted prototype, rock crystal, pearls, 6 x 22 cm designer's collection. 2010.

Generative tools (Jenn 3D algorithms for visualising complex mathematical geometries) can mimic the structuring of soap bubbles and provide a set of virtual geometric relations that can be explored and manipulated. Lynne was keen to build her own generative software program to provide her route to true creativity. She learnt Processing and Rhino3D with Grasshopper and her PhD research into a form of generative design called 'shape grammar' led to a deeper understanding of the conceptual process of this type of designing where the tools are the rules.

Throughout her work Lynne is astounded by the myriad possibilities that result from playfully pushing simple rules to their limits and manipulating algorithms. Surprising and serendipitous happenings, occurring either as emergent properties or caused by errors in the coding, often provide creative insights. Eventually a design is honed from these iterations and selected for making.

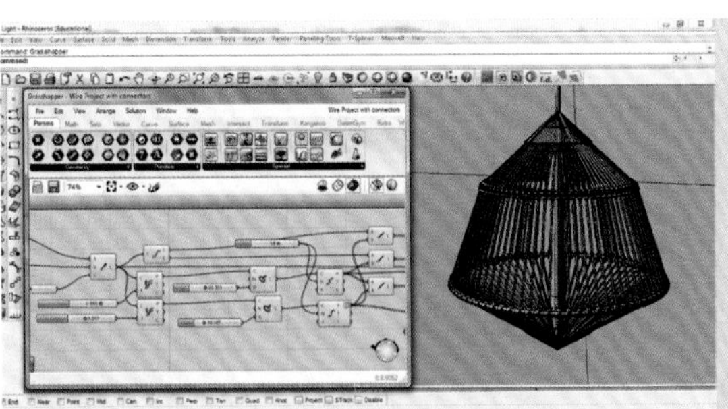

CUSTOMISABLE LAMP CONCEPT, Rhino3D and Grasshopper screenshot. 2011.

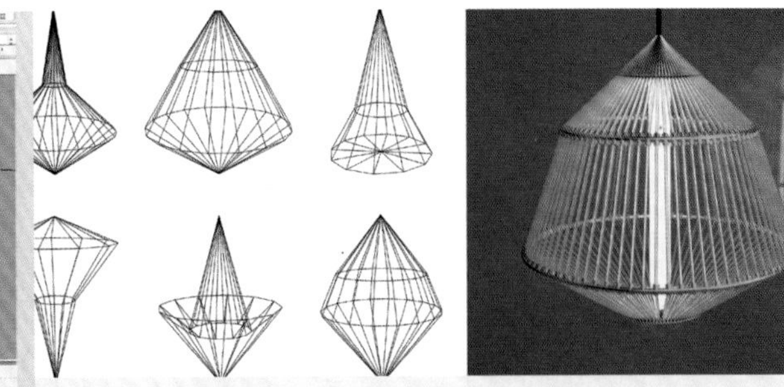

LIGHTING CONCEPT: explorations from Rhino3D. Render of lighting made from 3D printed gold plated steel and acrylic rod. 2011.

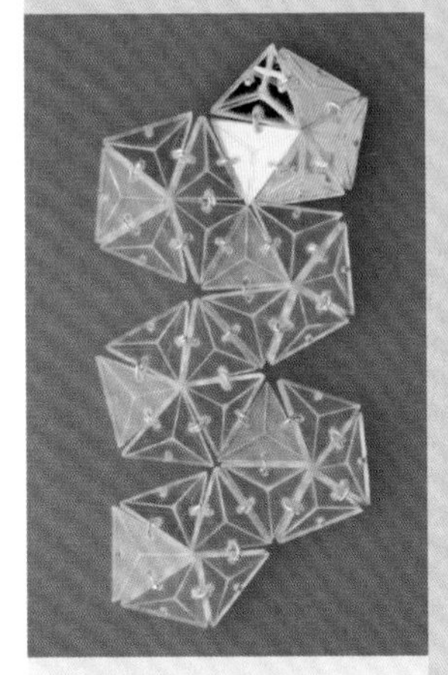

GEOMETRY STUDIES constructed by hand, laser-cut acrylic and steel jump rings, various sizes. 2011.

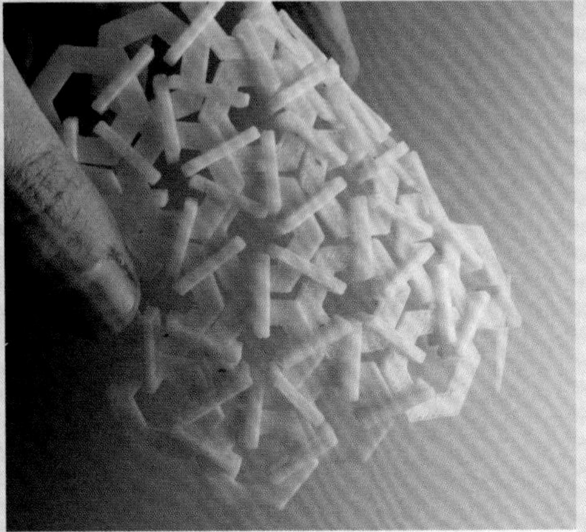

EARLY TESTS OF CHAINMAIL SYSTEMS, ABS 3D printed. 12 x 12 cm. 2011.

'PHASE' NECKPIECE, dyed 3D printed nylon and Swarovski crystals. The delicate graphic structure and form create shimmering Moiré interference patterns as the wearer moves around. 2012

Digital fabrication allows complex virtual forms to be made and electroformed, polished, and set with stones. Lynne's exploratory hands-on process directly feeds into and inspires new directions and highlights the importance of tangible models.

Laser cutting and 3D printing provide freedom to experiment – models reveal things that can't be known from a virtual model, such as how the object feels and moves around, crucial for informing practice and perfecting designs.

While using the tool built in Grasshopper she noticed that as she increased the number of 'spokes' on the designs, Moiré interference patterns appeared on the screen. She used this unintended, emergent phenomenon as the creative focus of her design process, informing the final outcomes for 3D printed work exhibited at the 3DPrintShow in London 2012.

Lynne's neckpiece illustrated above is part of her 'Phase' set that in 2013 won two gold awards in the UK's prestigious annual Goldsmiths' Craft and Design Council 'Craftsmanship & Design Awards'. In the 'Fashion and Conceptual Jewellery (Gallery Jewellery)' category the award is for both the design and the potential for reproduction in multiples, and this award highlights and recognises 3D printing as an inherently critical production technique for jewellery. Lynne's second award, 'Technological Innovation Award 3D', The Goldsmiths' Company Assay Special Award, 2013 (she was joint winner with Esteban Schunemann) celebrates the innovative use and contribution of technology for 3D jewellery.

This text is an edited version of Lynne MacLachlan's statement, sent to the author for this book.

PART 2

Essential: digital data

Creating data to access digital technologies

The first consideration for digitally creating is to find out which system and/or software package will deliver what you need and want. Usability is an issue for the individual to work out, as are justifying resources, time to learn and the possibilities of contracting out. Creating in 3D is more complicated than in 2D as software functionality can be 'hidden' behind an insensitive interface and the range of what is available can change rapidly. The cost of software is not a major issue as free and low-cost programs are available – they may have limited functionality but could offer just what is needed for the type of work you want to do. The main consideration is a program's learning curve and how long it takes to be productive.

'CULTURED ROSE' BY ALISON COUNSELL. CorelDraw for the artwork for photomask for etching: 'I use as much of the space as possible, so try ideas out if there is space (see lower left for 'nightlight' idea)'. Photo-etched and pierced stainless steel. Photographer: the artist.

RING BY DAVID POSTON. From hand-carved wax model, digitised by scanning in 3D (rendered here in Rhino), saved in .stl format and 3D printed in titanium. 2013. Photographer: the artist.

Designer makers want to use digital technologies in the same way as they pick up the tools and processes already in their 'toolkit'. They also want to understand how new technologies will add qualitative and quantitative value, which applications and combinations they should use, what kind of commitment they will have to make, what software courses are available, whether the software is easy to come back to and is it compatible with the actual process to be used, for example, laser cutting or 3D printing.

This book is for those designer makers who are considering using digital technologies but only if the benefits and advantages are such that they can justify taking time out from their work to do this. The key is knowing where to start and what questions to ask to get the information that is relevant to your own practice, as we all work differently, we use the same tools to do different tasks and we all have different perceptions and aspirations. As Zachary Eastwood-Bloom says:

there is an amazing amount of fluidity within using digital technologies in terms of inventiveness. Different makers bring different thought processes whilst using the same programs and systems, highlighting a massive creative potential.

Designer makers' tacit knowledge is based within different disciplines, materials, processes and tools. Their approaches to working can embrace 'what if' questioning, lateral thinking, time for reflection, etc., all of which means that supplying useful and helpful information can be complicated, particularly as technology and service provisions are evolving fast. I hope that by including input from other designer makers whose working practice already involves digital technologies, this book will provide information that is unbiased and down to earth.

Some of the work illustrated here is made by highly skilled professionals, which may seem intimidating. My route into digital technologies

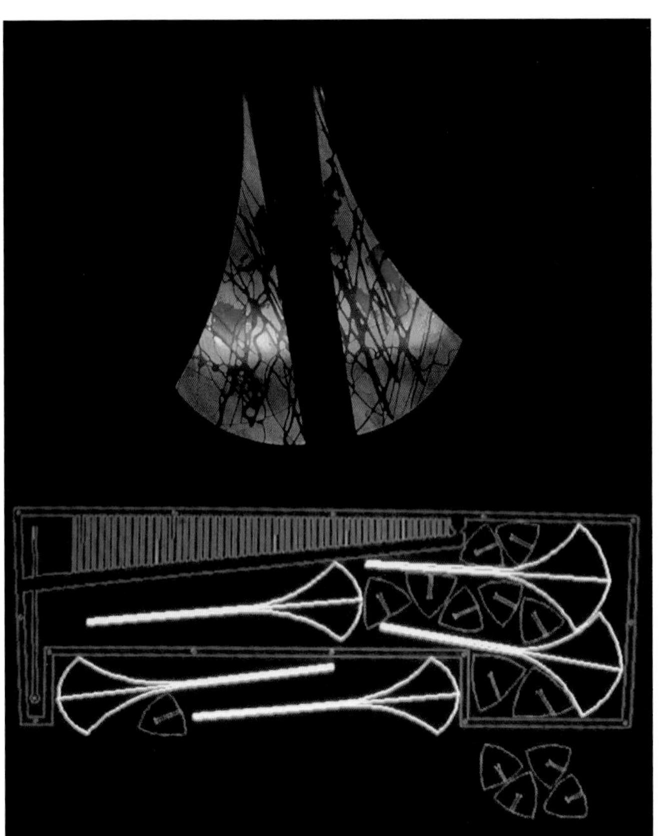

EARRINGS IN LASER CUT, ANODISED TITANIUM BY ANN MARIE SHILLITO. Learnt AutoCAD on the job to produce the digital designs for laser cutting earrings in titanium, and using up spare space within one section of trophy design for these. 1992. Photographer: the author.

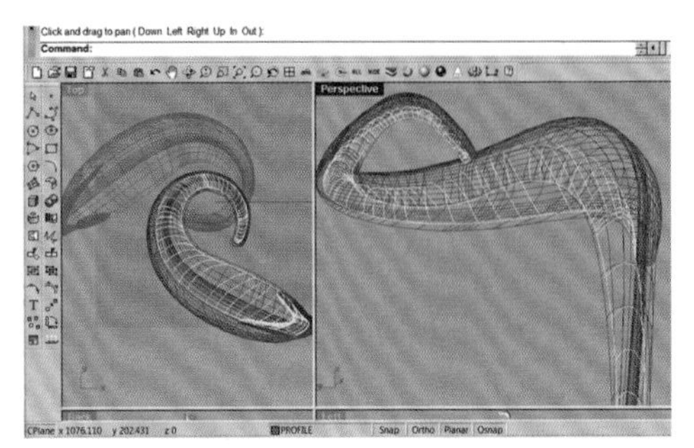

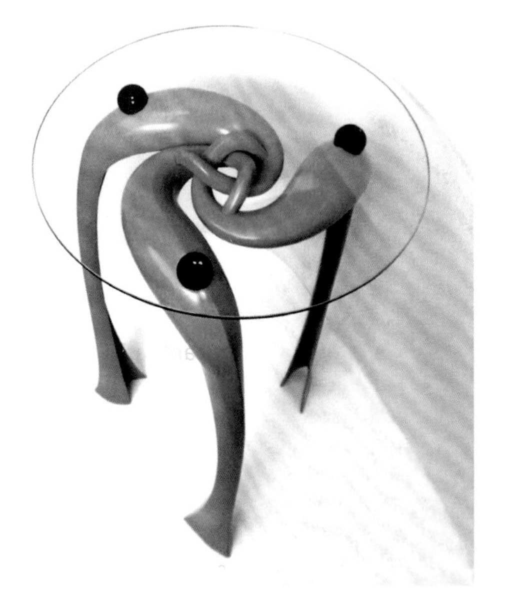

'COFFEE TABLE' BY ANN MARIE SHILLITO. Process: sketching, experimenting/designing in CAD (learning Rhino v3 and 3D digital design on the job), final design in .stl format, 3D printing in layered paper to test for fit, one leg 3D printed for mould, six legs vacuum cast in polyurethane for two sample tables. (Funding: Inches Carr Trust Award). 2000. Photographer: the author.

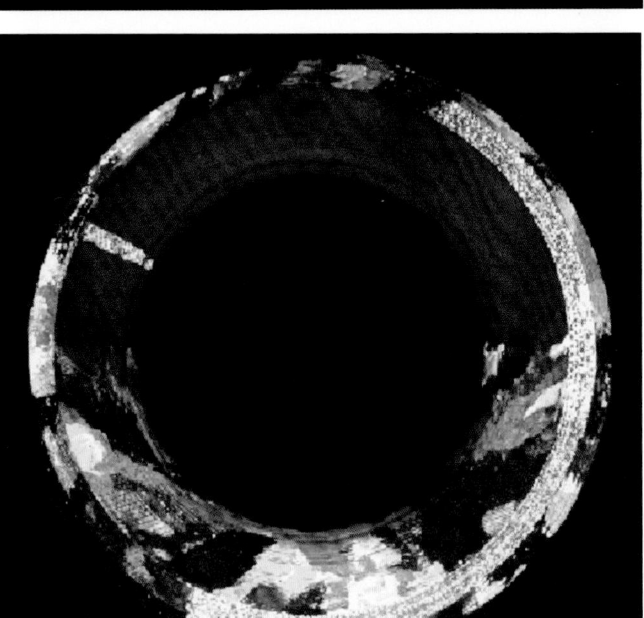

'VORTEX' BANGLE BY ANN MARIE SHILLITO. First piece designed using 3D CAD/3D modelling (TriSpectives). 3D printed in ABS, with three freely rotating rings. Painted with acrylics and finished with gold leaf. CALM Project. 1997. Photographer: the author.

should encourage even the most technophobic to feel less threatened and alienated, particularly with continuing improvements in usability and accessibility. In 1989 – spurred by economic necessity – I started using digital technologies by laser cutting my titanium and niobium jewellery range for serial production. In 1996/1997 I began modelling digitally in 3D to 3D-print directly wearable jewellery with no further processes such as casting or electroforming required. In his book, *Makers*, Chris Anderson says:

The biggest transformation is not in the way things are done, but in who's doing it. Once things can be done on regular computers, they can be done by anyone. And that's exactly what we're seeing happening now in manufacturing. (Anderson, 2012, p. 18)

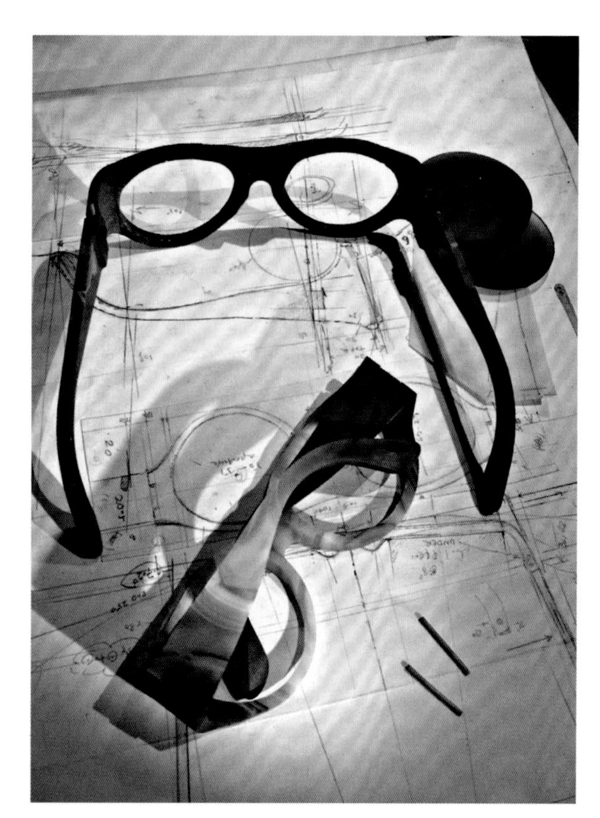

PAIR OF SPEC FRAMES BY BEN GASKELL. Made in semi-precious stone: 'It was really just a pencil and templates job. I did not use waterjet in the end, but if I had I would have wanted to "manualise" the control.' Photographer: the artist.

The best method for you

There are, however, times and tasks when non-digital methods and hand skills are still the better approach. Of course, virtual objects cannot be handled and this means that working digitally provides too few qualitative sensory experiences. Touching, feeling and seeing the material object is fundamental to its appeal, and this is especially true for beautifully hand-crafted work.

Ben Gaskell works with semi-precious stone. Each piece has unique and specific properties and flaws that only by using manually operated tools can he physically feel how to proceed. To be able to do the same work using digital technologies, as well as applying his tacit knowledge and know-how, Ben would need the highly-sensitive hapticated digital tools (using force feedback) that some surgeons are operating with today.

Different paradigms

Drawing with pencil and paper is not just the least expensive method of quickly capturing, visualising and iterating design ideas. Drawing is about thinking. Kurt Hanks writes:

Yes I had learned to draw; but more importantly I had learned to think... As my hand sketched the lines, my mind revealed a whole new method of thinking that I had not known before... What happened to my mind was much more important than the sketches I had produced. (Kurt Hank, *Rapid Viz*, quoted in *Pencils Don't Crash* by Ian Lambert and Richard Firth.)

Does drawing digitally, using features such as 'undo', remove such important impacts? Digital applications offer different and new paradigms. One is the chance to work in a 'risk-free environment'. Standard computer functions such as 'delete', 'undo/redo,' 'save', 'copy' and 'paste' support this idea, the most underrated probably being 'undo' and 'redo', which allow designer makers the opportunity to explore and push ideas 'over the edge'. 'Save as' removes the fear of losing track of interesting and promising ideas.

DIGITAL RENDER OF ALISSIA MELKA-TEICHROEW'S 'AMALIA' JEWELLERY COLLECTION (BLUE) in her 'Jointed' 3D printed series. 2012. Render by the artist.

Risk-free environment

The 'risk-free' environment is significant for 'prototyping'. Economically, prototyping in any form and degree is valuable for what is often referred to as the '1—100—1,000 rule': if it cost one to fix in the initial stages of the project, it will cost 100 times more to fix at the end of the project and up to 1,000 times more to fix once it is delivered. It is about 'fail fast, fail cheap'! Fearing failure, and thereby playing it safe, can stifle innovation, creativity and progress, so if a 'risk-free' digital environment can provide the platform to test ideas and objects, get feedback and discover that brilliant ideas are theoretically right but practically not quite right, or that you are on the wrong track, this is one sensible reason to embrace digital technologies.

In marketing, virtual rendered images of potential products can test marketable value and appeal before laying out financially for production tooling. If the response to the product is mostly negative, the digital design can be easily modified, customised or withdrawn entirely. Designs iteratively saved, incrementally changed, modified, retrieved, re-worked and developed into different versions for different purposes, can be rendered for presentation to different customers, commissions, exhibitions, retail opportunities, transmitted by email or via the 'Cloud' to collaborators, contractors, clients and friends. Images can be used for promoting practice through social networking sites such as Facebook, Pinterest. The list is endless and extending.

Another thought-provoking concept, with both pros and cons for the designer maker, is the 'gravity-free' virtual environment. For exploring ideas it is helpful for parts not to fall off and objects to tip over. Virtual objects, by default, have no mass and are unfettered by gravity. They can therefore float and pass through and into each other. These extraordinary properties

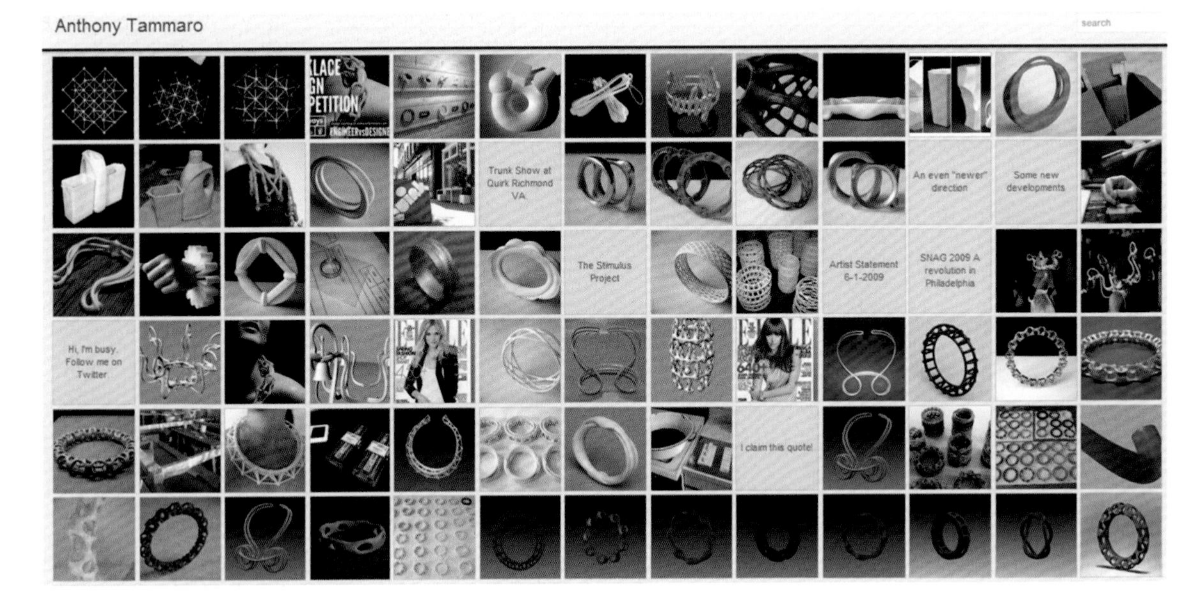

ANTHONY TAMMARO'S blog that informs his social network: images of new 3D printed work, statements and interactive links to exhibitions and conferences.

can be a distinct advantage if used with common sense, making sure that the objects designed are physically viable and makeable, that the bits that should be joined are joined, and that objects are topologically sound and will be stable under the force of gravity.

Which software?

Which software is best for me, for what I need to design and for the manufacturing process I will use? Can I justify the resources that will be needed to buy and learn at least one software package, which offers a sufficiently satisfying learning experience to keep me motivated, not frustrated, offers new possibilities and opportunities into which I am personally prepared to invest time, effort and money? These are just some of the questions to consider before purchasing any software packages. For examples of different software types see 'Do I need digital technologies' on pages 134–5. On pages 157–8 you will find details of websites and blogs such as Fabbaloo and up-to-date information and lists of the more common packages being used.

There is a significant range of software packages available for working two- and three-dimensionally. Some will only do 2D or only 3D, others will do both. Some packages are free and quite basic, other can be purchased for very little, but you can also spend thousands of pounds on one package. Some are very complex and over functioned for the needs of designer makers, so don't dismiss basic pacakges too quickly. We each have different requirements and different ways of working, and with new developments and methods for creating digital designs occurring at such a great rate, it is impossible for me to recommend specific products. To assist with your decision-making you will find throughout the book references and statements to particular software packages from designer makers, who give their reasons and choices of software in the context of their different disciplines and practice. In particular, chapters 4 and 5 have descriptions of software packages suitable for access to main manufacturing technologies such as laser cutting and 3D printing.

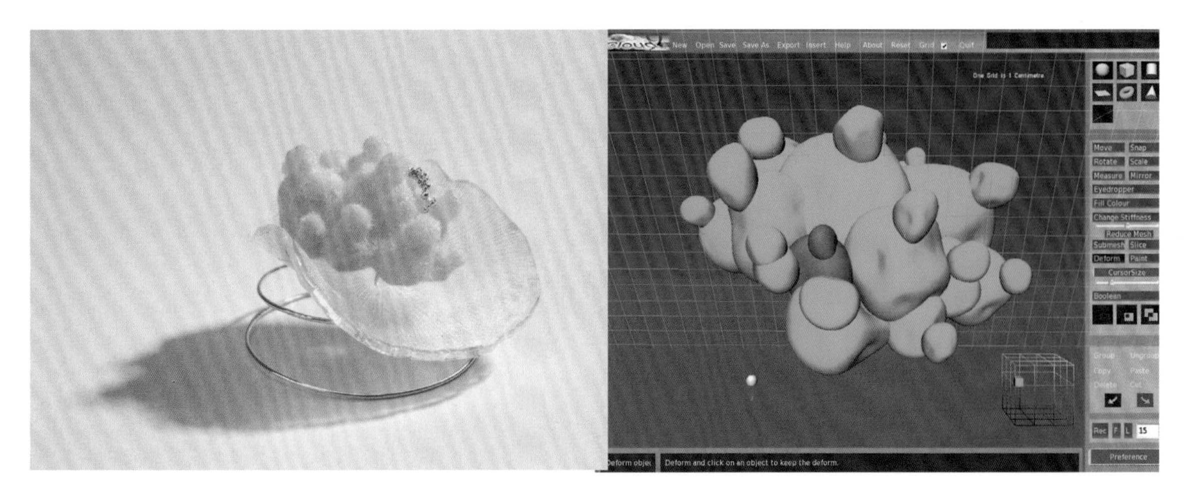

'SPRINGIN' BROOCH BY ELIZABETH ARMOUR. Inspiration: microscopic spore and fungi forms. Cloud9 haptic 3D modelling software ideal for organic forms. 3DPrinted in ABS plastic on UP3DPrinter. Now has own Cloud9 software. Brooch size: 3 x 3 x 3 cm. 2012. Screen capture: Anarkik3D. Brooch photographer: Malcolm Finnie.

Deciding which type of package to choose starts with knowing why you want to go digital, if you want to design from scratch or base digital work on existing designs (yours or from someone else) and which type of package you need: 2D, 3D or using maths/algorithms/scripting for generating form. There is software purely for graphic visualisation, for realistic rendering and for designing and modelling for digital production – these categories should be different enough to allow you to make a start. Artist Zachary Eastwood-Bloom says, 'I use the render package Maxwell, which means I can render my objects in whatever materials I am considering.'

'ECHO SHIFT' BY ZACHARY EASTWOOD-BLOOM, Cast bronze, linking programs, such as Terragen, Rhino, Illustrator, Maxwell and plotting software, and marrying these new technologies with historic processes that are thousands of years old. 15 x 65 cm. 2009. Photographer: Dominic Tschudin.

[By using a render package] I can visualise everything I want to make ... This is great for commissions and evaluating pieces. [It] appeals to my way of working.

Zachary Eastwood-Bloom, artist working in multi-materials

Important for the next stage of choosing your software is knowing which type of package will be best for you personally: some free software offers a pared-down introduction to a larger professional package (some with a vital function blocked); some professional applications are available on a short 'free trial' basis; some applications are stand alone, Internet/Cloud-based, and/or open-source developments. Upgrading can be expensive, although (oh sacrilege!) you don't have to upgrade. According to Dr Cathy Treadaway, 'software is ... someone else's logic and once you get he hang of it the next version is released ... more money and more to learn.' Before you upgrade, any impact on overall cost including learning time and cost of training has to be factored in.

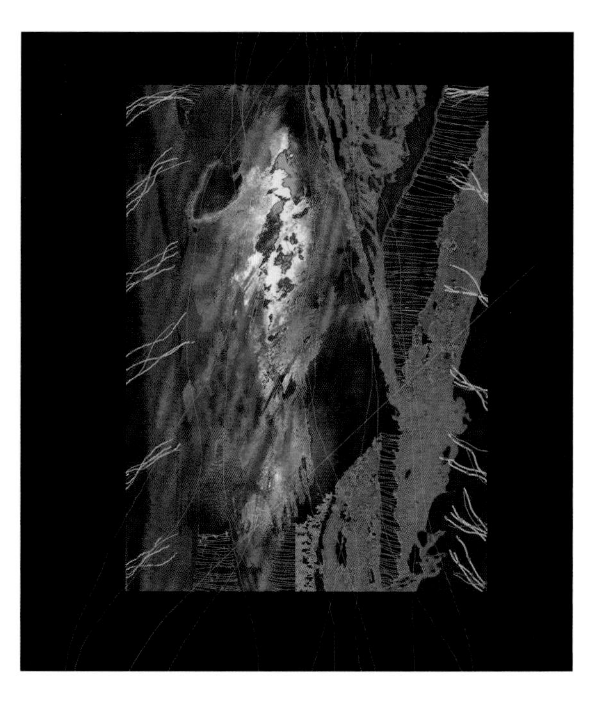

'BLACK FIREWORK' BY ALISON BELL. Artwork in Photoshop, printed on silk: integrating digital and traditional techniques. Photographer: the artist.

'CIRCLES IV' BY JENNY SMITH. Illustrator software used to create vector files to laser-cut and etched paper. 38 x 38 cm. 2009. Photographer: David Gough.

Two of the biggest concerns for anyone beginning to engage with new ways of working are: first, how can you achieve levels of competency in order to reap the benefits, and secondly, how can you mitigate the risks? It is not easy to take time away from hands-on making in the workshop, maintaining skills and the daily needs of your business in order to take on new skills, or to explore new ways of collaborative working and communicating in a different medium. Many designer makers feel they have little choice but to continue with conventional means of designing and making and relinquish advantages that digital technologies might provide. Lionel Dean, founder of FutureFactories and a big fan of 2D CAD has found, 'You can create the necessary 2D geometry in 3D packages but AUTOCAD does it so much better. Perhaps this is because I trained in old-school manual drafting?'

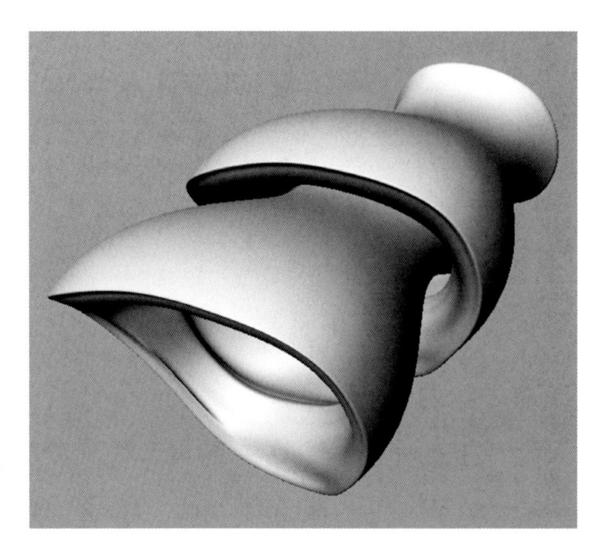

RENDER IN RHINO BY FARAH BANDOOKWALA. Design of articulated 'spine' parts for her 'Quiver' interactive sculpture (Jerwood Exhibition 2011)

[With] high-end hardware and software costs, new skills and training, more time is spent working around modelling software, compatibility, file size issues rather than on the design itself!

Lionel Dean on using computers

Is it necessary to learn to use every function well? Alison Bell, digital silk artist, chooses tools in Photoshop that are intentionally simple, such as the brush tool. She describes the brush tool as 'Just that, a tool like a brush, to be used by me' and says, 'I [don't] want to be transfixed by technology.' Artist Jenny Smith is only interested in gaining enough experience of her applications to enable her to prepare her own work and files and retain enough 'innocence' to exploit accidents and chance mishaps and to understand how these happened.

At college, Farah Bandookwala did a basic class in Rhino and then spent hours on her own building on what she had learned. What struck her was that she learnt the difficult way, making mistakes and having to redo things over and over, and that this process was actually very important because it allowed her to make up her own rules rather than having to follow anyone else's. Lionel Dean is breaking new ground all the time and does not expect to have all things well plotted, as no software designer is ever going to cover all eventualities, especially with users playing and making up their own rules.

Willingness to persevere

Human computer interfaces are often poor, not making full use of bodily interaction and skills.

Dr Cathy Treadaway,
Reader in creative practice

The complex interface of many graphics and computer-aided designing (CAD) packages is an obstacle to designer makers and is a result of programs being heavy on features and functions, many of which are not needed by most of us. Computers do not seem to be designed or developed for easy engagement with digital praxis, imposing an unfamiliar environment that reduces our ability to tap into or apply our tacit knowledge and hands-on skills built up through practical doing. However, designer makers' doggedness and willingness to persevere to find the right approach, tools and process for a job, are useful traits for overcoming digital obstacles and do predispose us as highly suitable users and 'subverters' of digital technology.

MICHAEL EDEN: THINGS MACHINES HAVE MADE...

Making is not only a fulfilment of needs, but of desires – a process whereby the mind, body and imagination are integrated in the practice of thought through action. 'Power of Making' exhibition catalogue, Martina Margetts, 2011.

'AMALTHEA' BY MICHAEL EDEN. Made by 3D printing a high quality nylon material with mineral soft coating. 20 x 55 x 24 cm. 2011. Leeds Museums and Galleries, Given by the Art Fund. Image courtesy of Adrian Sassoon.

As a maker, this statement defines Michael's practice and his good fortune to be working when there is a new creative momentum, brought about through technology and as part of a growing community of makers getting their hands dirty digitally, hacking the software, tinkering with the hardware and in doing so creating poetic and meaningful objects. He refers to himself as a maker in the sense of creator, happy to explore the overlapping 'grey' area between art, design and craft using whatever he needs to communicate an idea or story in three-dimensional form, be it a computer, potter's wheel, 3D printing machine or kiln; they are all tools requiring skill to do the job well.

'Unless artists can ... push and pervert their software far beyond its expected parameters, they must accept having their role as author/composer downgraded to performer.' Furniture maker, Fred Baier is cautious about the use of computers by makers. (Fred Baier, 2011)

Michael avoids being wowed by computing's fabulous box of tricks by becoming increasingly skilled in unconventional use of 3D software tools where there are endless possible actions which are risk-free. With a background working with clay before coming to computers, he applies the same sensibilities, starting by creating simple shapes. Challenges and risks begin to occur when he is improvising and using tools in ways that may not have been originally intended.

He discovered that digital technology involves a different way of thinking and problem-solving, waking up another part of his brain. 3D printing was exciting with the prospect of making the 'impossible' and his MPhil research project at

Michael Eden is a ceramicist who has fully immersed himself not only in discovering what digital technologies can do for him but also how he can fully exploit them and push boundaries. He discovered they involve a different way of thinking, a different way of problem-solving to ceramics, waking up another part of his brain!

Having come to the software after years spent giving life to clay, I am trying to apply the same sensibility to objects through the mouse and keyboard.

Michael Eden, ceramicist

A certain mindset is useful for maintaining fluency both in hand skills and digital skills. Adapting from a strong making preference can feel distinctly disempowering, as taking control and ownership of digital technologies can be elusive. Why is this? Historically, 2D and 3D design software programs have been developed by engineers for professional designers and engineers tied to using programs on a daily basis, week in week out. Sennett talks about the ten thousand hours 'it takes for complex skills to become so deeply engrained that these have become readily available, tacit knowledge' (Sennett, R. 2009. p. 172). Learning to master CAD is no different. CAD and graphics packages are complex and precision-based to meet the requirements of industry for production, whether

'THE WEDGWOODN'T TUREEN' BY MICHAEL EDEN. Made by 3D printing with non-fired ceramic coating. 2010. Crafts Council collection. Image courtesy of Adrian Sassoon.

'MAELSTROM VI' BY MICHAEL EDEN. Made by 3D printing. 2011. Image courtesy of Adrian Sassoon.

the Royal College of Art in 2006 was the ideal opportunity to start exploring CAD and bring the two worlds (and the two parts of his brain) together.

Although the virtual is no replacement for the actual, CAD proved very useful for iterations of ideas, providing enough visual information to determine whether geometry and proportion would work. Bouncing between traditional processes and CAD, translating virtual into actual, he concluded that their relationship was healthy: each has its place and there is a reason to use both. His final, fully digital MPhil piece was 'Wedgwoodn't Tureen', an iconic object and a narrative homage to Josiah Wedgwood (the famous potter who lived and worked during the first Industrial Revolution), redesigned with a delicate pierced surface inspired by bone.

Impossible to produce using conventional industrial ceramic techniques, as gravity, centrifugal force and the material qualities of clay limit the forms one can throw on a wheel, the 'Wedgwoodn't Tureen' was designed on Rhino 3D and FreeForm software and demonstrates the removal of these constraints for potential to create previously impossible forms that can creatively communicate new ideas. It was 3D printed on a ZCorp machine, the first to use a non-fired ceramic material developed by the French company Axiatec, which Michael adapted to closely resemble Wedgwood Black Basalt.

For his pieces 'Maelstrom' and 'Vortex' Michael adapted Rhinoscript to digitally emulate the technique of coiled pottery construction by manipulating a selected group of cylinders to form one cylinder, stretched and shaped until the desired form was created. Experimenting with heat-treated ZCorp plaster the 3D-printed pieces have been successfully glazed and fired: collaboration with the University of Washington's Soldner Laboratory in Seattle resulted in biscuit-fired 3D-printed ceramic pieces, glazed and fired in Michael's studio, and with the Digital Manufacturing Centre at the Bartlett School of Architecture, UCL, London, successful tests of 3D printed ceramics were achieved.

This text is an edited version of Michael Eden's statement, sent to the author for this book.

kettle, car, film, advert or building, as these demand precision and prescribed management pipelines to deliver goods *en masse*. This and other CAD functions such as verifying and dimensioning parts, file transfer formats that conform to industry standards, are important, necessary and useful, but do not fit easily into designer maker's practice where subjectively exploring and developing concepts and contexts are prime, not acquiescing to industrial constraints.

Designers like Michael, balanced between 'right-brain' and 'left-brain' thinking, more easily incorporate CAD's engineer-focused structure into their practice. Their learning curve is less steep and when they return to pick up CAD again, they do so swiftly and effectively. For many of us CAD's prescriptiveness can produce continual cognitive disruptions destroying 'flow', an immersive condition and deeply qualitative experience that designer makers cherish and work so hard to achieve within CAD's complexity, leading to frustration and a sense of exclusion.

In conclusion, there are two choices here: designer makers can adapt to CAD's more prescriptive way of working, or they can search out those applications that better suit their way of thinking and doing. In this context business guru Seth Godin's advice is pertinent: 'The opportunity of our time is to discard what you think you know and instead learn what you need to learn.' (Seth Godin, p. 13)

4 Applying 2D digital technologies

FRUIT BOWL BY ANN MARIE SHILLITO. Designed using 2D CAD package for data for controlling laser cutting. 1992. Photographer: the author.

Knowledge of digital technology is no longer an option but a necessity.
— Inge Panneels

Just as the constraints and affordances (that is to say, advantageous properties and qualities) of materials and making by hand influence the design process, the different types of software have a bearing on all stages of design. Rigidly separating the design and making processes into different sectors does not represent how designer makers work and how work is made.

I define '2D digital technology' first as graphical (images/visual representations such as photographs and drawings made up of pixels and vectors) and secondly as 2D CAD (geometry/mathematics based, rendered graphically as drawn lines and shapes). Mac computers dominate graphics and so most Mac software is pixel and vector based, ideal for printing and for the less heavy-duty profilers, cutters and lasers. PCs dominate CAD. The more heavy-duty processes such as waterjet, routing and powerful laser cutters use industry standard CAD formats and examples of software include Form Z, Rhino, AutoCAD. Thus, this chapter is roughly split into the lighter weight 2D technologies (lower powered machines and vector-based instruction) and the heavy duty 2D technologies (higher powered machines and CAD-based instruction), as the approach to each can be very different.

Pixels and vectors

A pixel is a square dot that is the smallest element of a digital picture. Its name comes from 'pix' meaning 'picture' and 'el' meaning 'element'. Hundreds of thousands of pixels can make up the image generally referred to as 'bitmap' or 'raster'.

Each pixel has different attributes and properties such as colour (red, green or blue) and tone, blending visually together to make up the image. The colour depth, or resolution, of each pixel, depends on the rendering power quality of the graphics card, and as the range can be in the millions – more than the eye can discern – a great amount of detail can be presented. An image created on a standard computer display screen (as opposed to a photograph) is constrained to 786,432 pixels in total (1,024 pixels across by 768 pixels down).

'NECKPIECE #1'. DRAPE SERIES BY STEPHEN BOTTOMLEY. Silver electroformed onto laser cut acrylic. Inspired by original Fortuny geometric textile motifs and 'softness' of fabric, vector based design from textile scan manipulated in CoralDraw to intentionally avoid perfection and precision commonly associated with CAD. Diametre 410 mm. 2007. Photographer: Charles Colquhoun.

The disadvantages of pixels compared to vector formatting (see below) are that each must be stored separately as it is rendered individually, meaning pixel-based images can take up greater disk space than vector-based rendering. Jagged edges can also be a visual problem when scaled up.

Adobe's Photoshop is a popular pixel-based program that includes standard file formats (jpg, tiff, psd, etc.) compatible with processes such as printing in black/white and colour.

Vector-based software generates images of lines and individual objects (primitives) constructed on sufficient instructions (i.e. x,y coordinates for two end-points on a 2D plane, for direction and size) for each to have its own attributes and characteristics that can be individually changed, making editing easy.

Large vector graphic drawings and pictures don't need lots of disk space; lines can be zoomed to extremes and scaled-up vector images remain sharp. Vectors can be very accurate, making this system highly suitable for plotting and cutting as the coordinates are converted into toolpaths to control machinery. Adobe's Illustrator and CorelDRAW packages are well known vector-based systems.

Images in digital format are multi-purpose, and growing compatibility between different formats means that a colour photograph of pixels can be rendered into black/white so the vectors can be extracted as outlines for use with different marking, engraving, piercing and cutting processes and manufacturing systems (printing, photo etching and piercing, plotters, laser, waterjet, electrode, blade, hot wire or router cutting).

'SUMMER1' DESIGN BY ALISON COUNSELL, drawn on CorelDraw as vectors. Render by A. Counsell.

Capturing sketches and images

Intricately incising and piercing very thin and fragile bone china forms was part of Beate Gegenwart's vocabulary of drawing in her early days as a practising ceramicist. She used Walter Benjamin's writing as a starting point, in particular his 'Arcades Project', which brought much new thinking and required her to:

adapt [her] drawing practice to allow new influences to shape the work ... The initial drawing process is a "rehearsal" for the permanent marks to be cut; areas are removed by the laser, describing space, [the] lines creating shadows on the wall behind forming the "double", connecting artwork, wall and panel. (Also see pages 74–5.)

'ARCADES 2' (DETAIL) BY BEATE GEGENWART. Stainless steel, laser cut, felt. 65 x 65 cm. 2011. Photographer: Chris Stock.

◀ **'LIVERPOOL MAP' BY INGE PANNEELS AND JEFFREY SARMIENTO.** Waterjet fused and printed glass, 6 m x 2.25 m x 33 cm x 5 cm, Museum of Liverpool. 2010. Photographer: Simon Bruntnell at www.northlightphotography.co.uk

▶ **'WOMAN + LOCH 1 STAGE 2'. DIGITAL IMAGE FOR PRINTING ON SILK BY ALISON BELL.** Discovering digital photography, developing simple skills in Photoshop to digitally collage precise photographic images with new layers of digital process. Photographer: the artist

Combinations of tool and process in design may include 'paper and pencil', and 2D scanning is invaluable for transferring photographs or an original artwork's quality and nuances into a pixel format. For one of her laser-cut pieces, Jenny Smith scanned her handwritten, notes on how to use the laser-cutting machine and used Illustrator to convert it to vectors, painstakingly dragging words, 'cutting' and rejoining lines, maintaining the element of mystery and active engagement for viewers as they try to decipher the written meaning.

Inge Panneels, a glass artist and academic describes how her making processes 'rely on a combination of traditional hand drawing and CAD drawing techniques: images are drawn using a Wacom pen and pad'. These input devices simulate paper and pencil more closely than the standard WIMP interface. More recent digital 'mark-making' applications use touch screens that make finger or stylus sketching and drawing very appealing. They are easy to use, the effects are stunning and the images easily printed out. The advantage these processes have over pencil-and-paper is that drawings are in a digital format (pixels and vectors) and transferrable to other programs such as Adobe's Photoshop and Illustrator to use higher level functions.

2D digital data for production

There is printing and there is printing.

Alison Bell uses Photoshop to create the artwork and then prints directly onto silk using a state-of-the-art Mimake wide-format digital printer at Glasgow School of Art's Centre for Advanced Textiles (CAT). This gives her the freedom to explore and experiment with colours and designs without limitation.

Katie Bunnell prints colour and pattern onto ceramics. The designs on her ceramic plates involve an extra process whereby the pattern and colours, created using 2D scanning, photography, Photoshop and Illustrator are digitally printed as 'decals'. Decals are a ceramic substrate, applied to the plates and then fired for permanence.

Alison Counsell prints resist onto steel for photo-etching and piercing for formed and folded steel, combining what she loves: 'hand-drawn imagery with the crisp accuracy of the industrial photo-etched stainless steel.' With CorelDraw she can 'maintain hand-drawn quality' using a mouse and a freehand drawing tool, and she can also use

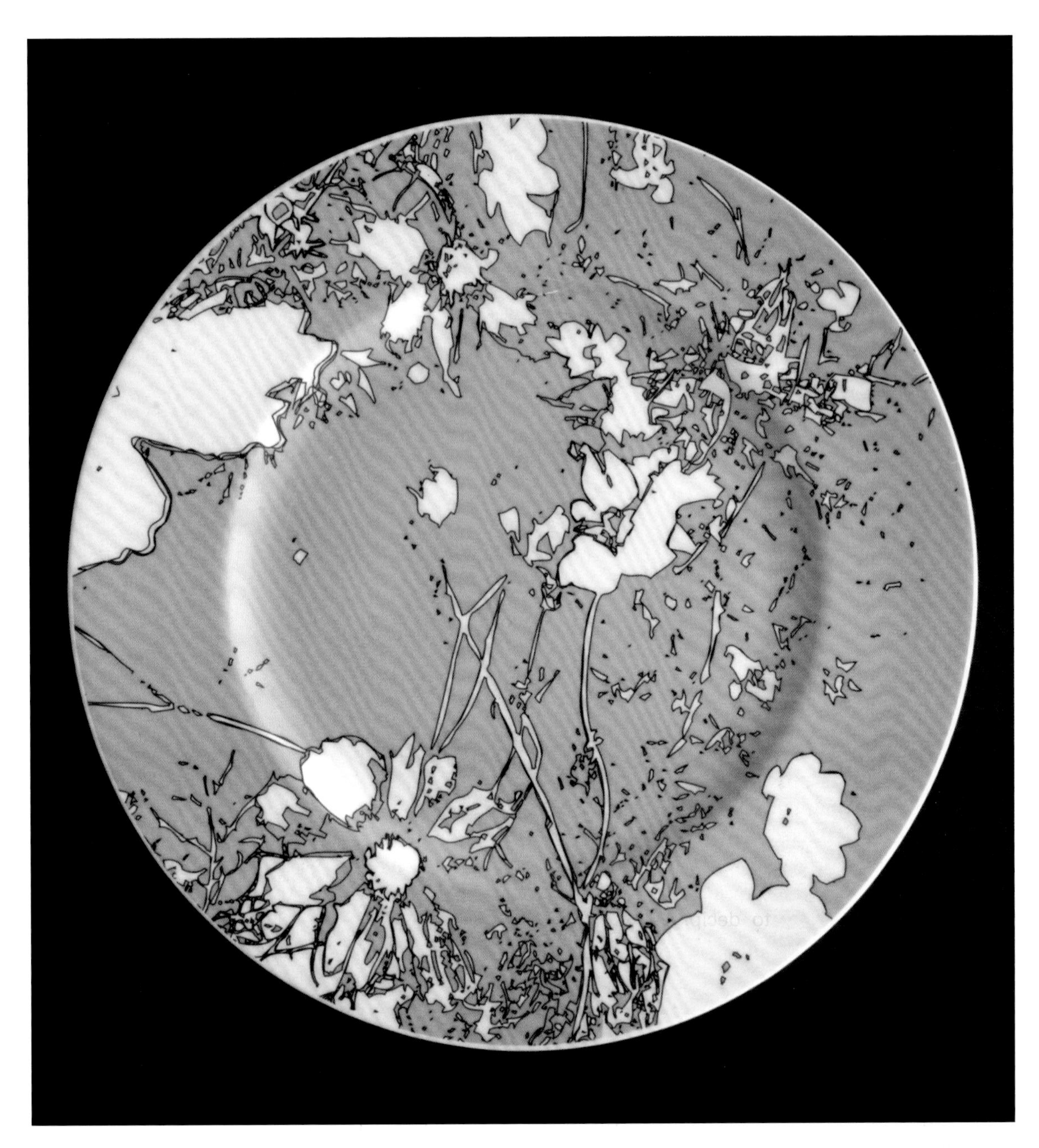

'DIGITAL FLORA' BY KATIE BUNNELL.
Digital ceramic project: online interaction with designer-makers collaborating with clients to customise or prototype designs for tableware. 2006. Photographer: the artist.

ART WORK FOR 'CULTURED ROSE' AND 'BED OF ROSES' BY ALISON COUNSELL. Design drawn in CorelDraw to produce two aligned sheets of pattern for masks for photo-etching and piercing. Stainless steel. Digital image and photograph: the artist.

accurate 'layer' and template facilities for efficient coordination of the drawings for both sides of the sheet metal to be etched. Importantly, she found a great photo etching company that has been open to some of her experimentation.

Some designer makers see being once removed, with no autonomy over process, as a disadvantage. There is no opportunity to say 'stop there' or 'what if' and thereby discover new possibilities at each stage. Vector-based software and equipment do have more flexibility and capacity for artistic sabotage than CAD and heavy industry systems. Where degrees of accuracy are not critical and the software and the hardware are to hand, designer makers will use and subvert tools, equipment and machinery for that 'what if' exploration. By having artwork, studio materials, technologies for formatting, making and finishing in close juxtaposition, flitting between them is easy. There is time to reflect and think. There is more chance of spotting something different. At these times serendipity and new and innovative events happen.

Jenny Smith now has her own laser cutter so that she can personally manage the entire process. She recognises that '"accidents" within the digital process are becoming a central part of my methodology and allow what could be considered quite a rigid part of the technical process to open up layers of further creative potential.' (*Digital Calligraphy*, 2009) Getting hands-on access to equipment, materials and information really does facilitate experimentation. More institutions are providing open access to their workshops with short courses at all levels, and specialist practitioners like Jenny Smith, offer 'beginner' and 'master' classes (see Jenny Smith at Edinburgh Laser Cutting Studio). FabLabs are a growing global network of 'fabrication laboratories' that were started at Massachusetts Institute of Technology's Media Lab. Their open and unrestricted approach pitches art and technology together for great collaborations and enables 'toe dipping' to try out a process, create a 'bread and butter' line, change direction, find ways to differentiate products and move one's practice up several notches.

From digital drawing to tangible work

One of the most straightforward ways to learn to use digital technologies is through service bureaux and hubs such as Ponoko and Formulator. There is a trade-off though, as procedure and how artwork is presented are tightly stipulated, and their range of materials is restricted to standardised sizes and thicknesses. Their websites have extensive and useful guides with step-by-step instructions, the types of programs (e.g. Illustrator, CorelDraw) that will produce compatible file types (e.g. vectors), file formats (.ai, .eps, .dxf, .dwg etc.), templates and size for artwork/design (1:1 scale, border widths) to fit machine-bed dimensions, line weights and types of cutting and how to mark in art work (red lines for laser cut, black for raster engraving, blue for vector engraving and scoring, greyscale image files as .jpeg, .tiff or .png for engraving), presentation of text, naming files, how long it will take to get a quote and how long the job will take.

Ponoko offer bespoke, easy-to-use 2D design software for laser cutting and routing, a design starter kit covering the costs involved (making, material and shipping), tips and tricks to keep these as low as possible, as well as more specific information such as how much material the laser burns away and therefore how much space you need to leave between units for the kerf. Just making sure a design is clean and the lines consistent with the guidelines (no duplicate lines and continuous lines with no breaks) can save a lot of hassle on both sides to guarantee a successful job.

In one of his research papers, Gilbert Riedelbauch proposes that uploading your digital file to an online fabricator like Ponoko is a new kind of collaboration as the maker's design visualised on their screen is the 'visual programming' that communicates with, and directly controls, how the fabricator's machine produces the part.

'RADIAL' BY OTTO GUNTHER. Designed using Adobe Illustrator CS5. Laser-etched onto wood, cherry and birch veneer, through Ponoko. Hand inked (acrylic ink) and finished. 39.4 x 80 x 5.7 cm. 2011. Photographer: the artist.

Lightweight 2D production

Vectors control and direct cutting and marking the material lying flat on the machine bed using a router or laser beam for melting, burning or vaporising. With prices falling rapidly, 'lightweight' laser cutting is exceptionally easy to access, opening up opportunities to try it out, be more experimental, prototype fast and 'dirty', work more commercially and move efficiently and effectively between one-off tests, bespoke commissions and small production runs. The advantages of this are being able to design resourcefully in series, in multiples of the same shape, to exact specifications, with amazing levels of fine-cut detail and in materials ranging from paper, card, leather, fabrics, rubber, felt, acrylics, melamine, cork, wax, wood and composites to thin sheet metals. It is no wonder laser cutting has found wide application amongst designers and artists: product, textile, fashion, jewellery, graphics, illustrations, artwork, presentations, promotions, shop window displays, commercial product ranges, bespoke work and even food (laser cut nori for designer sushi).

SOME COMPARISONS BETWEEN 2D CUTTING AND MARKING METHODS

'ARCADE 3' BY BEATE GEGENWART. Mild steel, laser cut, vitreous enamel and laser engraved. 58 x 22cm. 2011. Photographer: Aled Hughes.

'SKETCHCHAIR' FROM GREG SAUL AND TIAGO RORKE. Prototype chair being CNC machined from Valchromat, coloured fibreboard at FabLab EDP, Lisbon, Portugal. 2011. Photographer: Tiago Rorke.

What industry rejects as dimensional inaccuracy or unwanted effects can be a positive advantage and an attractive, distinctive feature, for instance, the rough, wave-like pattern on the cut surface, the scorched edge of laser-cut paper and wood or the polished edge a laser leaves on acrylic. Another example of this is to be found in inlaying. When waterjet cutting thick material, if the jet exits at a different angle from its angle of entry, an essential lightly tapering edge is created. CNC cutting can give far more flexibility for play and experimentation for stunning effects and is generally more accessible.

Laser and waterjet compared with physical CNC machining

Sheets for laser and waterjet cutting require minimal jigging and fastening – set-up is faster, especially for parts that are difficult to hold or easily damaged using mechanical clamps. Easy set-up makes smaller production runs and one-off pieces more practical, particularly where high accuracy and repeatability are required.

Tighter, sharper angles can be laser or waterjet cut as CNC milling can only be as fine as the diameter of the smallest available tool. Laser cutting is not limited or affected by the hardness of a material – high-carbon steels can be cut as easily as standard mild steel. Lasers cut more accurately than either waterjet or CNC cutters and, with laser beams measured in hundredths of a millimeter in diameter, higher levels of detailing are achieved. Sheet metal cut with waterjet is not prone to the buckling that the concentrated heat of a laser beam can cause.

Only a limited range of materials can be cut economically with waterjet cutting, i.e. the long cutting time of tool steel can outweigh the advantages. With other materials, the higher cost of laser and waterjet cutting when compared to CNC routing is mitigated at post-processing stage where mechanical cutting leaves rougher edges that require additional finishing. The hard-to-get-at edges of a laser cut require no further work.

The advantages of laser cutting also apply to laser marking, which produces clear, burr-free and permanent results – with no mechanical deformations. Even difficult and inaccessible surfaces can be marked with ease.

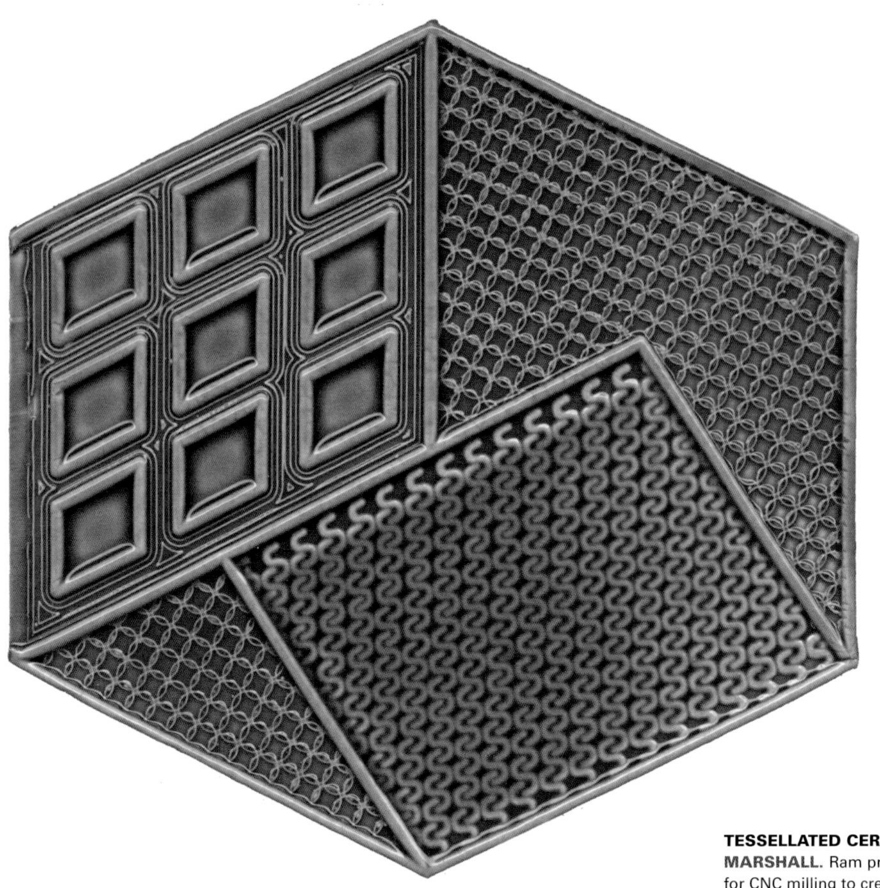

TESSELLATED CERAMIC TILE BY JUSTIN MARSHALL. Ram pressed ceramic. Tool path for CNC milling to create master pattern: vector file from Illustrator. 26 x 24 cm (each tile). 2000. Photographer: the artist.

With laser etching/engraving there are two techniques for surface effects. The first uses pre-coated metal, which has been surface treated: anodised, lacquered, powder coated or spray veneered. The laser removes the coating to reveal the metal below. The other 'etches' the bare metal surface using a slow but high-power laser, which alters the surface structure of the metal without removing any metal.

Laser-cut perforations and computer digitally controlled (CNC) milled grooves are used to constrain folding. Gilbert Riedelbauch uses lines etched and engraved in malleable, flexible materials. Wonderful, undulating surfaces are created with curved lines. Small-scale CNC milling machines engrave, score, groove and cut soft materials on the flat in 2D (X-axis moves left and right, the Y-axis moves back and forth) and will carve in 2½ D with the Z-axis controlling depth (with no undercuts possible).

Justin Marshall developed a tessellated tile using Adobe Illustrator, importing the vector file into software to create the tool paths for a desktop CNC milling machine. This cut the master patterns in thin milling board from which the silicon rubber moulds were made for the ceramic tiles.

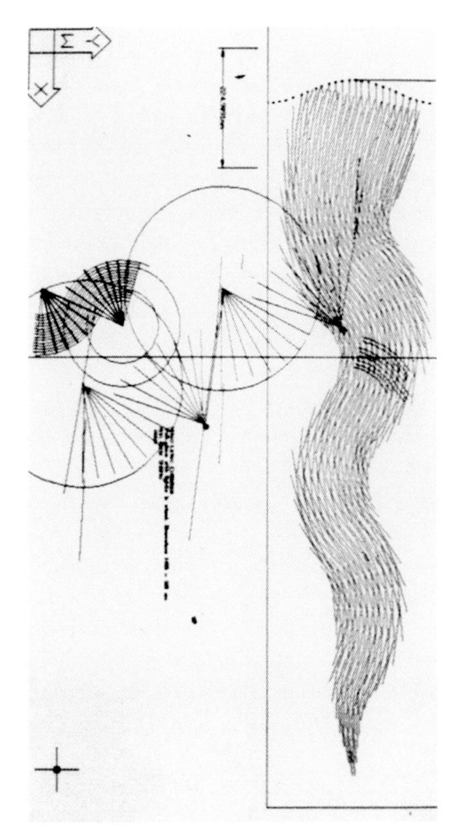

DESIGN FOR WALL PANEL (SCREEN CAPTURE OF DESIGN IN CAD) AND FINISHED PANEL WITH DIOCROIC SPOT LIGHTS, IN LASER-CUT STAINLESS STEEL BY ANN MARIE SHILLITO. Designed in 2D format, using construction lines and layers, saved in .dxf file transfer format for laser cutting. 1999. Photograph of panel: Shannon Tofts.

Heavy-duty 2D production

Higher-power milling, laser and waterjet machines for marking and/or cutting hard, chunky or exotic materials, metals, ceramic, glass, composites are governed by the industry's requirements for absolute control and tight tolerances to produce precision-based widgets and use CAD generated data. The praxis of direct access for experimenting is near impossible as the process is twice-removed: machines and machining are fully automated with the process controlled throughout by technicians.

Inge Panneels is a freelance glassmaker and an academic at the University of Sunderland, where Dr Vanessa Cutler did her pioneering PhD research into glass waterjet cutting. For Inge, the most exciting aspect of this technology is cutting glass shapes that are impossible to make by any other means to open up a new, less inhibited strand in her work.

The precision cutting of the waterjet machine allows an intricacy of glass parts that was hitherto impossible by traditional means. The excitement of this new language is tempered by the fact that the processes involved were not acquired intuitively but relied heavily on input from others; from learning to draw digitally using a pen and graphics tablet, relying on others to make the digital files in the correct format and writing the Lantek code for the waterjet machine. (Inge Panneels, 2012)

Thus the nuances she wants as an artist continue to elude her. The advantages are sufficient for her to persevere with CAD and collaborating, as she does find the use of CAD inspiring and invaluable regarding economy of time and scale, allowing designs to be thought through and visualised quickly, placed within the context of space, amended and scaled, with fast extrapolations from the design printed off on a large scale for life-size construction drawings. But she does require

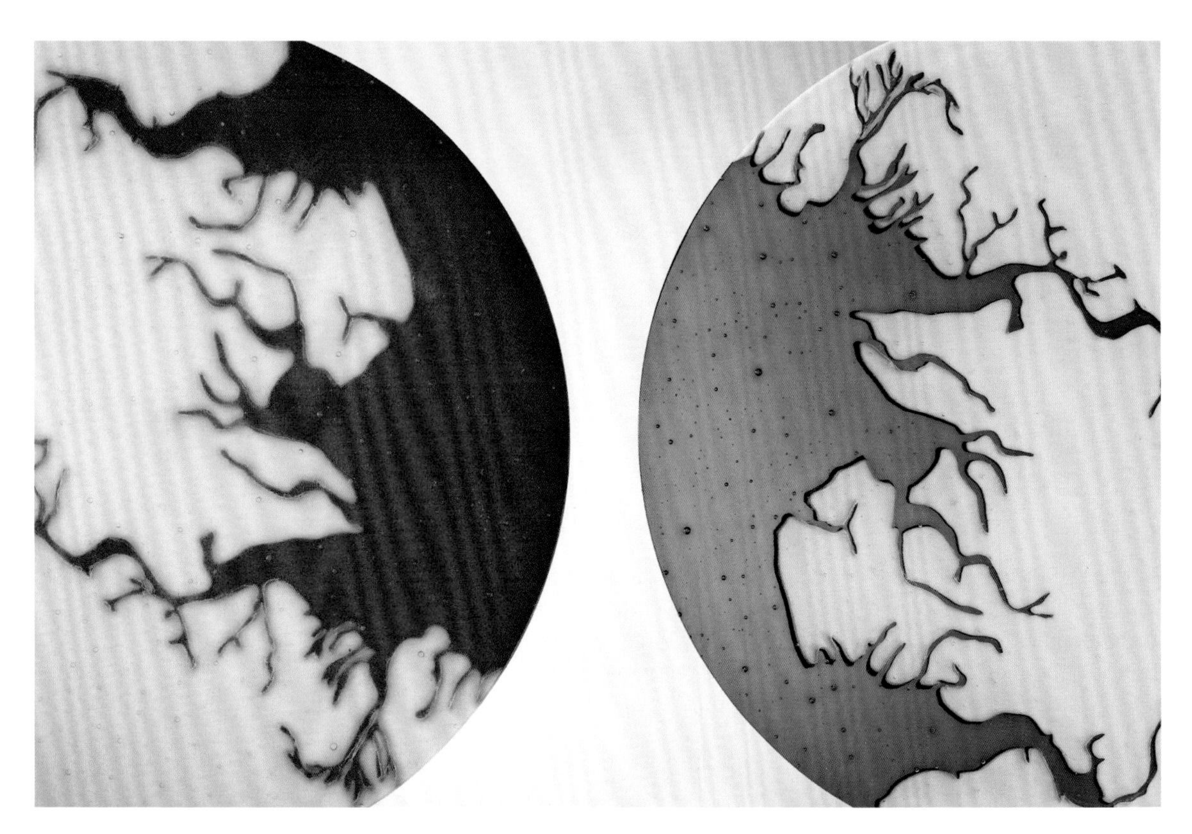

'MICRO MACRO' WALL PANELS (DETAIL) BY INGE PANNEELS. Waterjet fused glass, 2 x 28 cm diameter. Private. 2011.
Photographer: Kevin Greenfield.

technical support at most stages of production to ensure digital files between Photoshop, Illustrator, AutoCad, Archicad, and InDesign are converted into the right formats and, for waterjet cutting, translated into the programming control language (WaterJet Sweden's LANTEK). Inge relishes the potential for being able to design in a different way, whether in Rhino, Illustrator or Photoshop, and to be in control without requiring a 'translator'. As Inge says, 'The future is bright and exciting!'

CAD

CAD is engineering based and so provides file formats (such as generic IGES and STEP) which hold dense unambiguous data for precise manufacturing. With more open systems, designers and architects are able to import/export designs between different CAD packages with a range of file formats and use the best program for the specific task in hand. In practice, access to different packages is needed as well as the time and resources to learn to use them. The design transfer process isn't always a smooth one as products can have a historical legacy of different construction protocols and architecture. So between pixels and vectors and 2D CAD, between 2D and 3D, voxels (3D pixels), mesh-based surface and solid modelling, parametric modelling, and new hybrids, actually translating data cleanly from one system to another is a complex process. Enterprising companies have stepped in and develop middleware and plugins to bridge a gap or two. Greater computing power and clever algorithms have pushed much of this complexity into the background for more seamless transitions between software and methods of production.

The laser cutting company I used for titanium and niobium jewellery units still prefer the standard transfer formats of .dwg and .dxf, both of which hold less dense data, describing only the relevant parts of the design required for cutting purposes.

CAD/CAM

CAD/CAM stands for computer-aided design/ computer-aided machining. Routing is straight-forward profiling, cutting flat sheet with the path of the high-speed rotating cutting tool directed and driven from CAD data. The machine bed and sheet move in two axes (X and Y) and a router head moves down in increments in the Z-axis.

'DRAPERIES' OBJECT BY DANIEL WIDRIG. Daniel Widrig is a designer and proficient user of digital technologies across the 2D/3D spectrum. The system he uses depends on the project or commission and material used, whether CNC machining polystyrene or polyurethane. Polystyrene, CNC machined, 50 x 100 x 10 cm. 2009. Photographer: the artist.

Computer-aided milling can be 2½D. As with 2D machining, all levels must be reachable straight down from above – there can be no overhanging elements. Multiple passes using different kinds of cutting tools and inclined toolpaths allow highly complex curves to be perfectly carved out of different materials from foam to wood to metals.

'I B POP' RED-WHITE CHAIR **BY BLUE MARMALADE.** Interlocking forms make the flexible recycled polypropylene rigid. 2006. Photographer: Magnus Bjerk.

Blue Marmalade designs and produces contemporary furniture and the final production of units procedes on automated presses, the 'production tool' from a CAD drawing made by various fully automated CNC operations. Their freeform designs are assembled by hand from the shapes stamped out from their own blend of recycled polypropylene sheet material (with additives for some unique properties).

To maintain absolute control they use AutoCAD in the early design stages to iterate and model quickly in 2D the three-dimensional forms that are in their heads, while also working to tight

'TEMPEST' PENDANT SHADE BY DESIGN COMPANY BLUE MARMALADE. Flowing interlocking forms hide the light source from view but are bright and functional. Shade made from 100% recycled plastic for zero landfill policy. 2011. Photographer: Chris Lomas.

'TERRAIN'. FOLDING CLOCK BY GORDON BURNETT. CNC machined and anodised aluminium. H 250 x W 50 x L 325 mm. 1994. Owner: maker. Photographer: Stuart Johnston.

production tolerances. The digital data enables an easy jump from hands-on test pieces to CNC cutting for all subsequent development, as once designs are finalised they move on from expensive CNC prototypes to production tooling using CAD data to design the stamping tools in house to their specifications: they work to much smaller waste margins than most companies and all the waste that cannot be designed out is recycled back into material for new products in a closed loop.

Gordon Burnett, a silversmith using various software and CAD/CAM tools, embraces the interrelationship with, and the interdependence of, materials and tools. Martin Heidegger identified this in 1977: 'materials and tools work together in partnership with the act of making and the maker in a relationship of co-responsibility.' (M. Heidegger, 1977) Gordon learned numerical machining code (G-Code) to control a three-axis CNC milling machine and then 'Quicksurf' software on a Mac to create more ambitious forms. He found that:

Occasionally, however, what was shown on screen was then altered by the software making adjustments for machining, resulting in unimagined but welcome subtle surface patterns. 'Is diving in the digital ocean fit for practice?'

For most of us any level of control relies on understanding enough about how machining is controlled. This is an essential aid to designing, not only for what can be done but also to make sure that as we design our 'drawing' is coherent, consistent and clear. For engraving, routing and laser-cutting this means that the digital design is flat on one plane, clean with no duplicate, overlapping lines, certainly no extraneous floating, unattached bits and no gaps between the ends of supposedly continuous lines. Gaps can be so miniscule that they are only visible by zooming in and in and in and in. Any 'break' is in the code and affects the smooth running of the cutting process as the machine stops and then starts again. To avoid wasting time searching for 'gaps' I quickly learned how to use CAD's 'snap' function, which not only prevents 'gaps' but also holds my line flat on one plane (I was using a 3D CAD package). It saves a lot of hassle and misinterpretations to fix methodically as you work. With laser-cutting, 'gaps' are particular messy as with each start-up there is a surge of power which creates a larger starting hole and a significant burn mark. To artists, this burn effect can be an interesting feature, but if you are not present to say 'I like that, I want to keep that', technical staff will 'correct' this along with all other aberrations and you pay for their time.

Waterjet cutting

The software requirements and file formats for waterjet cutting are much the same as for industrial strength laser-cutting as they work in the same way: waterjet uses a highly focused and pressurised stream of water instead. Although waterjet is slower than laser-cutting, it delivers a higher quality, less-detailed cut. Unlike laser-cutting, no heat is generated and this is important where heat may distort and/or change the properties of the material, such as annealing tool steel or colouring the edges of titanium, or expansion cracking fragile materials such as glass, ceramic and stone. Waterjet will cut through materials up to 200 mm thick and can cut materials that laser cannot cut well (rubber, linoleum, felt) or at all. There is also very little material wasted (small kerf width) with waterjet, and precision parts produced have no burr or rough edges, eliminating the need for finishing processes such as sanding and grinding.

'ARCADE 1' BY BEATE GEGENWART. Stainless steel, waterjet cut, vitreous enamel. 60 x 60 cm. 2011. Photographer: Aled Hughes.

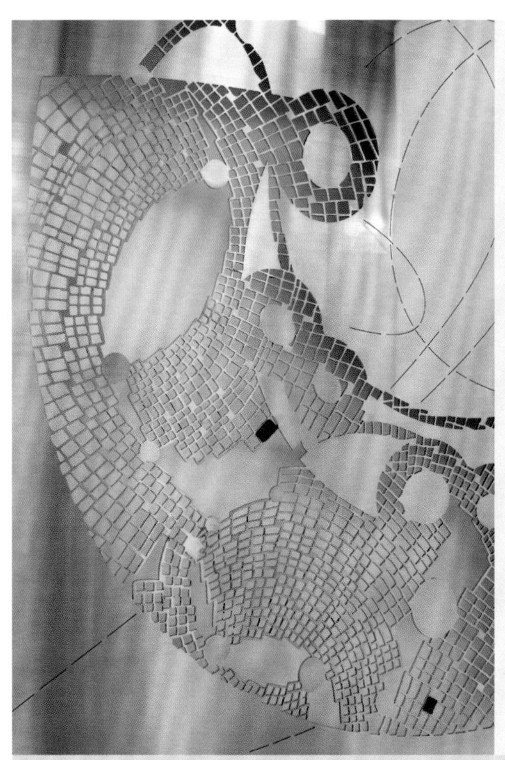

'PANEL 2' (DETAIL) BY BEATE GEGENWART. Stainless steel, laser cut, felt. Photographer: Chris Stock.

■ BEATE GEGENWART: DRAWING PROCESS AS A 'REHEARSAL' FOR THE PERMANENT MARKS TO BE CUT

Beate Gegenwart is an artist with a focus on visual computing, digital textile techniques and laser cutting. She has adapted her drawing practice to allow new influences to shape her work. Each piece is built 'using layer upon layer of drawn imagery'. She begins on paper, 'drawing lines to form net-like structures with empty spaces, holes. The lines became the space itself, conveying a sense of dynamic movement.' With Adobe programs now so overwhelmingly rich in possibilities, every piece takes advantage of Photoshop's sophisticated masking possibilities and Illustrator's drawing tools. When the work is complete, Illustrator is used to create the vector paths for cutting.

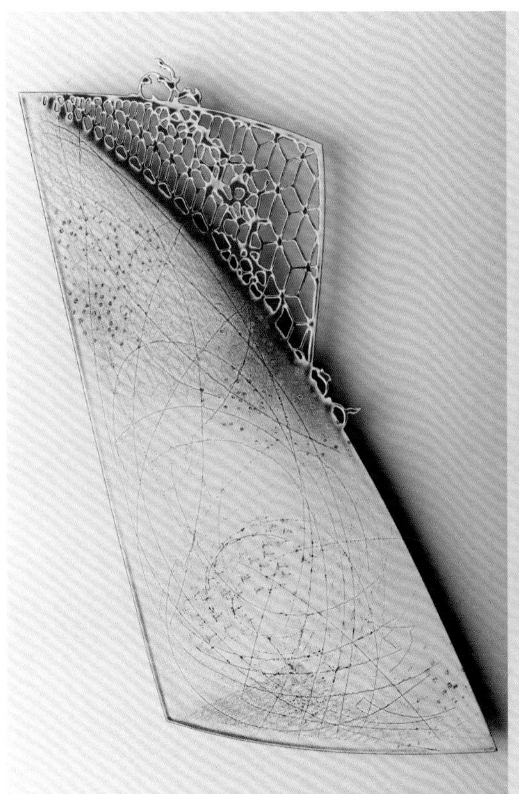

◀ **'ARCADE 2' BY BEATE GEGENWART.** Mild steel, laser cut, vitreous enamel, laser engraved. 58 x 22 cm. 2011. Photographer: Aled Hughes.

▲ **'ARCADES 4' (DETAIL) BY BEATE GEGENWART.** Stainless steel, laser cut, felt. 60 x 60 cm. 2011. Photographer: Chris Stock.

Each metal panel has its 'companion' in paper, cut in minute detail. The panels 'echo Walter Benjamin's tiniest notebooks filled with frenzied ... thoughts and collections of objects'. Cutting, incising and piercing have been part of Beate's vocabulary of drawing for a very long time, and as a practicing ceramicist she 'intricately incised and pierced very thin and fragile bone china forms, the incisions, lines and holes creating energetic movements around and across the forms'.

Her 'Panels' are either laser or waterjet cut, a cold process that removes any risk of warping, though the quality is not as fine or as sophisticated as a laser is on stainless steel. The act of laser-cutting is the more extreme as the heat slightly deforms the metal, creating physical tensions in strong contrast to the original intricate making, and she has to accept the 'non-perfect': every piece is different. However, lasers can cut lines and marks finer and smaller than she would ever be able to accomplish by hand.

'After this, the 'hands-on craft' begins: the metal is polished; vitreous enamel [applied, using] various techniques in selected areas. [Then] the work is fired in a large kiln and this is where there is an element of "chance". The artwork is fired and re-fired several times, the handmade mark is unpredictable and intimate and every batch of steel seems to differ. The oxidised metal is subsequently polished back to silver, the enamel is ground and stoned, and sometimes laser engraved, continuing the layering in the work, some digital, somehand, back and forth with lots of hand-finishing involved.'

'The greatest barrier and frustration is that during the "industrial" cutting part I have to rely on others. I would love to be able to do the cutting myself and to be able to research the parameters in depth, but unfortunately the equipment is prohibitively expensive and in terms of the waterjet cutter, the use is complex and difficult.'

This text is an edited version of Beate Gegenwart's statement, sent to the author for this book.

'ESTUARY' BY INGE PANNEELS. Waterjet cut and fused glass. 5 x 25 cm diameter. Private owner. 2011. Photographer: Shannon Tofts.

Access to and technical support in waterjet is via two routes: through companies providing the service to the industry, therefore catering for the requirements of industry, and through technical workshops in colleges and universities where the boundaries can be pushed. When online companies like Ponoko offer waterjet as a service, demand will increase as prices drop. Meanwhile, Charlotte De Syllas, acknowledged as one of the finest artist-jewellers working in Britain today, is experimenting with various digital technologies, including water-jet cutting, at Metropolitan Works in London to pursue her own research and develop her gemstone carving. The potential that waterjet holds to block out and pierce shapes in the chalcedony family and jades for inlays and inlaying was intensely tested on a black agate necklace.

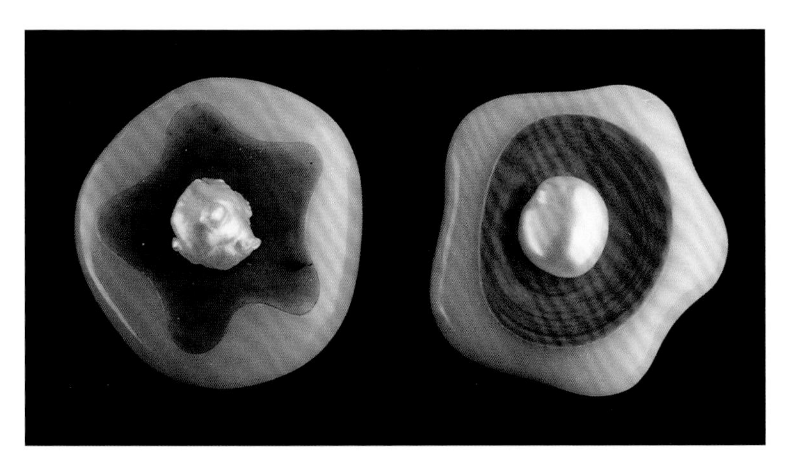

EARRINGS BY CHARLOTTE DE SYLLAS. Inlaying using white and green jades, and with coral. Waterjet cutting at Metropolitan Works unsuccessful so earrings are hand work. 2010. Photographer: Jasper Vaughan.

The stress that the cutting puts the stone through needs to be sorted [by using and adjusting a] variable pressure nozzle [to] enable pressure to be lowered for piercing and then raised for cutting. This has apparently proved very effective when cutting brittle materials such as glass and stone etc. The main weakness is when a hole is cut into a stone to cut out a shape from the inside of a piece. The force in one place can crack the stone but the travelling cut is less prone to breakage. A good quality chalcedony worked well for Ben Gaskell and I had no trouble with Cacholong. Black chalcedony is not so strong. (Charlotte De Syllas, 2010)

Ben Gaskell also works in semi-precious stone. Yet the agate spectacles he made were a 'pencil and templates job' and not designed or made using any digital technology save the convenience of digital cameras/scanners, used incidentally, and 3-axis CNC drive without direct computer control for some milling and drilling of the stone hinges. He reverted his manual-wheel drive method because with hand feeding for the drilling and milling he could feel (and hear – and sometimes smell) the danger point when the cutting is too hard. He says,

Every raw piece [of precious stone] is unique or at least rare and expensive and you soon realise that making these things by hand-controlled machines is not just better (more easily modified as you go along), less risky of breakages, but actually often quicker. (Ben Gaskell, email correspondence, 2010)

Ben would use waterjet but he would want to 'manualise' the control in order to feel and understand what is happening. The only way he and Charlotte achieve success is by tapping into their unique skills and knowledge gained from working long and closely by hand with different stones and their properties: it is nearly impossible to articulate or transfer tacit knowledge of what the danger point in cutting a particular stone feels like, sounds like or smells like.

Digital technologies are marred for many designer makers by this lack of sensory feedback, although haptic (virtual touch) systems are available for designing and 3D modelling and for hi-tech tasks such as robotic surgery. Once machining is enhanced with highly sophisticated sensors to give sensory feedback, designer makers will be able to apply a more hands-on approach to working with these technologies.

Other ways of making

Sarah Silve has a background in silversmithing and developed a laser-forming technology using CAD/CAM for automating the scoring of metal for folded objects and for producing repeatable and more affordable metal products. Her research at the Liverpool Laser Group from 1997 explored the capability of laser forming for silversmithing and the relationship between and the resulting forms from using heating strategies with 2D CAD patterns. She set up a laser facility at Brunel University in 2001 to continue her research and the take up and application of the process by other practitioners.

▸ 'X-TRACK@' VASE BY SARAH SILVE. G-code programming. Aluminium. Laser formed tube. 1999. Photographer: the artist.

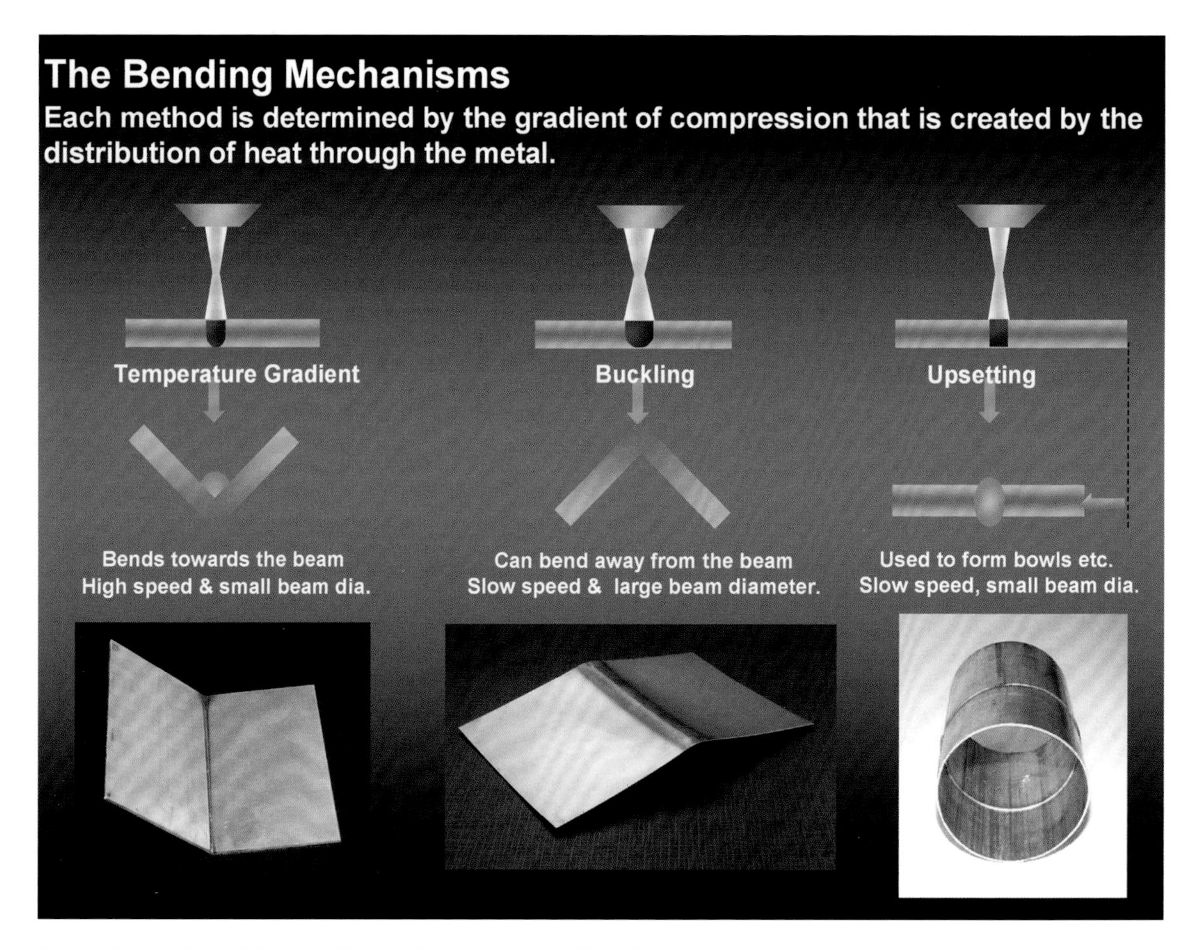

SARAH SILVE'S LASER FORMING MECHANISMS FOR SHEET FORMING. Diagram from the artist's paper, 'Lasers: forming a relationship in the making' presented at Cutting Edge: Lasers and Creativity' Symposium at Loughborough University, 2009.

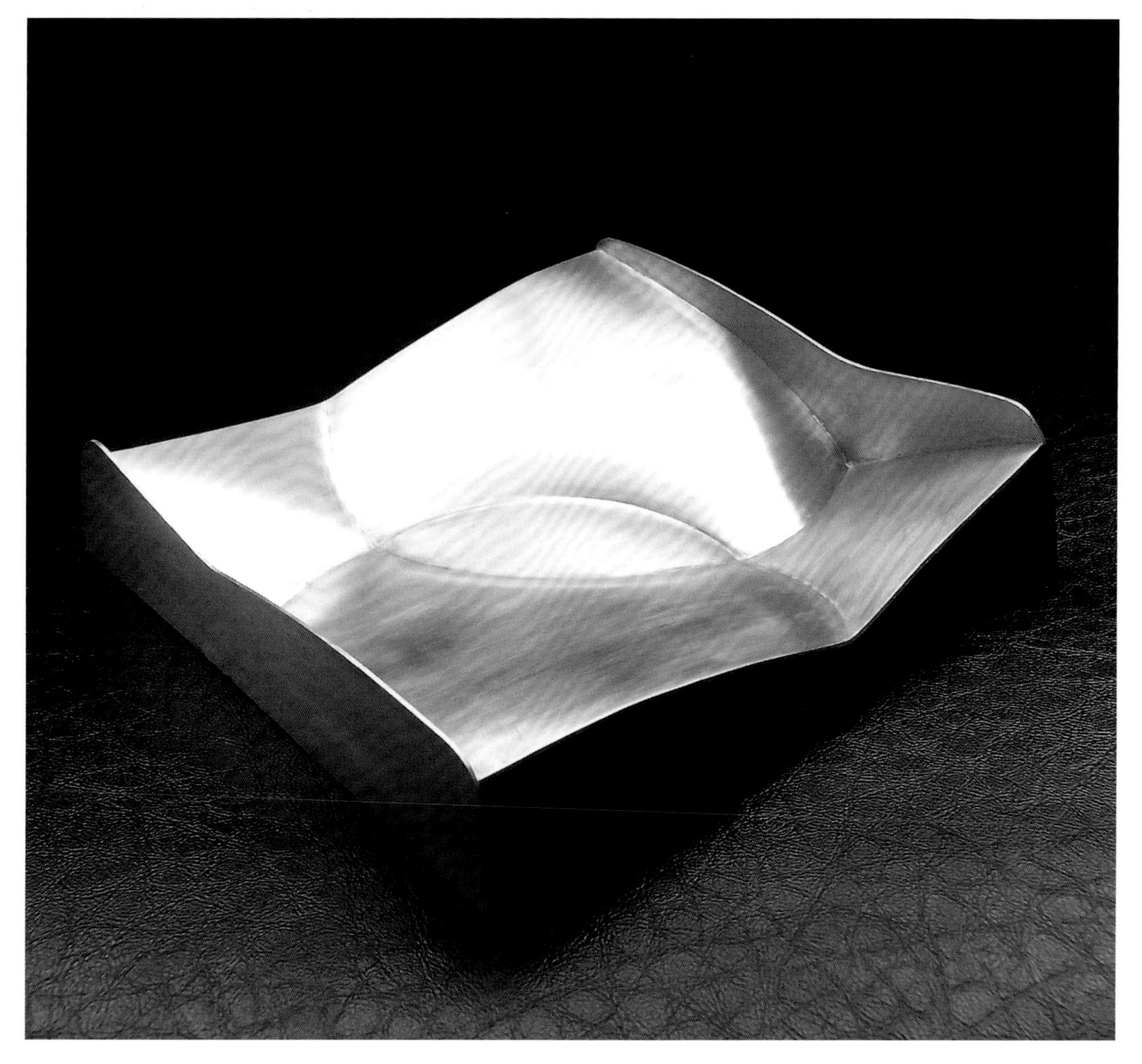

ELARCA DISH BY SARAH SILVE. BodCAD-CAM. Silver. Laser-formed and fabricated. 2005. Photographer: the artist.

The impressiveness of heavy-duty technologies can cause us to dismiss or undervalue less 'sexy' technical systems for production, for instance, the craft of embroidery. Banks of sophisticated CNC embroidery sewing machines are the mainstay for many companies who use automated processes for stitching digitised pattern into fabric. The maverick potential of such machines in the hands of designer makers is massive. For example, Nuno Textiles in Japan, founded by Junichi Arai and Reiko Sudo, uses CNC embroider to great effect in some of their cloth (nuno is Japanese for 'cloth'), purposefully employing unexpected combinations of materials and unusual treatments, for them any process (metal sputtering, blowtorching, rust dyeing, three-dimensional weaving) is fair game for experimentation.

I mention Nuno as I admire and respect their ethos: Nuno hold to the Buddhist belief of living lightly in the world and combine exploration of new materials and synthetics with digital technologies with respect. They also support traditional aesthetics and hand crafting techniques to make these Japanese treasures more accessible to textile lovers worldwide. They use machine embroidery to great effect to create a range of delightful textiles by reusing small offcuts and

WATCHSPRING BY NUNO TEXTILES. Cotton (80%) and polyester (20%). Using a chemical lace process, Nuno use a computer-controlled embroidery machine to embroider onto a water-soluble base fabric that is then dissolved away to create this coiled-steel watchspring design. 2009. Photographer: Daiju Goto.

STALAGMITE BY NUNO TEXTILES. Polyester (100%). Embroidered with a special 'steering wheel' computer-controlled embroidery machine, such as used for *uchikake* wedding kimono, the design based on calcified rock formations. 2005. Photographer: Sue McNab.

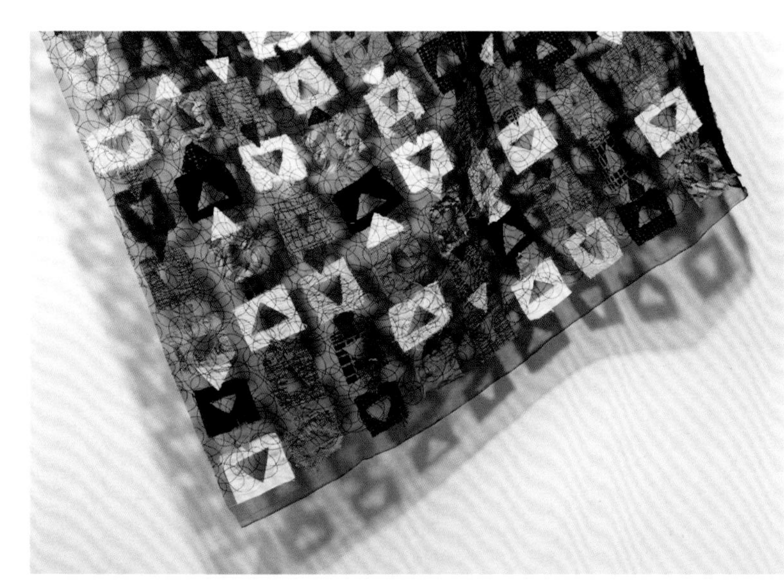

NUNO KASANE (TSUNAGI PATCHWORK SERIES) BY NUNO TEXTILES. Silk (100%). Their 'stock' of leftover fabric scraps are first ironed, each piece is then cut to size and basted by hand onto silk organdie before the fabric is finished and sewn by computer-controlled embroidery machine. 2005. Photographer: Sue McNab.

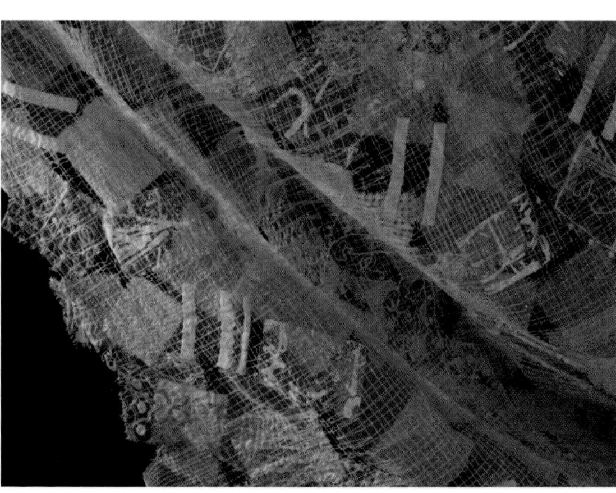

TUSGIHAGI – VARIOUS REMITTANCE OF NUNO TEXTILES (1997) AND NUNO TSUNAGI – SILK (2004), TSUNAGI PATCHWORK SERIES BY NUNO TEXTILES. 'Chips' of various different fabrics, patched together by hand before the fabric is finished, sewn by computer-controlled embroidery machine. Photographer: Sue McNab.

scraps, recycling polyester and nylon and reducing waste. Their CNC embroidered textiles were in their ZukoZuko exhibition at Ruthin Craft Centre and at the Dovecot in Edinburgh in 2012.

We do small-lot machine-made industrial textiles with a design edge. Since our founding in 1984, we have always worked within Japan because of the wealth of traditional weaving and dyeing skills that still exists among the many brilliant – and often underemployed – craftspersons and artisans thoughout the country. Without the aid of their knowledge and intuition, we simply could not realise our textiles. (Reiko Sudo, Abstract for Seminar at Designskolen Kolding, Denmark in 2009)

From hi-tech to lo-tech

Another digital system for producing test pieces and finished work is the 2D plotter cutters that cut out profiles and shapes in thin, non-metallic sheet materials like paper, card and vinyl using a small knife automated similarly to laser and waterjet cutters. 2D plotter cutters are available 'off-the-shelf' and are inexpensive enough to justify a place in workshops and FabLabs (MAKLAB has a vinyl cutter) for prototyping, and making finished pieces and installations. The SketchChair team (who prototyped their Antler Chair at FabLab EDP in Lisbon), used a cutter to make miniature chairs to test their system – as Ponoko recommends for reducing hassle and expense:

Print out your design on paper first. You could consider this a free and instant first prototype. It's the ideal way to spot sizing errors, see whether you've made holes big enough, and get a feel for what your final result will look like. (Josh J. www.ponoko.com)

In the early 1990s I used a plotter to draw the profiles of 2D outlines created in CAD (AutoCAD and Rhino), then I either cut out the paper units by hand or glued the units/sheet onto metal and pierced them out for early stage prototyping. The drawn lines were consistent, making the plotter very effective for creating artwork from which to produce the photo resist masks for etching/piercing silver and steel and for anodising patterns in titanium which match the profiles for laser

'ANTLER CHAIR'. SCALE PAPER PROTOTYPES AT 1:7, BY GREG SAUL AND TIAGO RORKE. Various paper iterations using SketchChair app and testing design and construction using 2D plotter cutter. Diatom Studio, London. 2011. Photographer: Greg Saul.

cutting. Inge Panneels uses her digital imagery to prepare files for screen-printed stencils used for sandblasting glass.

Scripting as a creative digital process

Some designers have long customised their CAD's user interface (UI) to automate routine and repetitive functions by using scripting code (programming language). In Rhino, for example, a button in the toolbar can be customised by holding the [shift] key and right-clicking it once to bring up the 'Edit Toolbar Button' dialogue box. Growing numbers of designer makers are really getting into scripting as another method for designing digitally (Gordon Burnett says that he 'began by learning to enter lines of numerical code in Heidenhain machine language directly into a three-axis computer numerically controlled [CNC] milling machine') combining this with more traditional tools and processes to extend their practice and facilitate new ways of thinking and doing. In CAD, it is data that describes and mathematically determines the 'patterns' of shapes. By changing even one parameter new patterns or forms are created, and software modified through scripting offers a range of speculations that are outside the manufacturers' intentions.

Grasshopper is a graphical algorithm editor tightly integrated with Rhino's 3D modelling tools. It requires no knowledge of programming or scripting, but still allows designers to develop form generation algorithms without writing code.

'100:TRIANGLE STOOL' AND 'YELLOW STOOL1' BY RICCARDO BOVO. Laser-cut plywood and cable ties. 40 x 40 x 40 cm. Artist's private collection. 2010. Photographer: the artist.

Processing is a software 'sketchbook' for learning the fundamentals of computer programming within a visual context and is so straightforward that tens of thousands of people are learning it to create images, animations, and interactions using it for playing, exploring, prototyping, and for generating finished professional work.

Using Grasshopper, Riccardo Bovo created his laser-cut 'One Hundred Triangle stool'.

Each stool is designed with a unique series of random triangles generated by an algorithm. Then the wood is cut using a laser and all the triangles are joined together with cable ties. The result is a light-weight stool with an innovative soft structure. The algorithm, created with Grasshopper, receives as input a seed number (for convenience I use the serial number of the stool), which is used to generate random values, [i.e.] a subdivision value used to determine the number of 'cells' that compose the stool, and a set of guide-surfaces that are used for growing the structure derived from the random values and the subdivision numbers. As an outcome, the algorithm produces a 3D file showing the resulting shape of the process and a 2D vector file that consists of the outline and number of each triangle carefully flat packed and ready to be fed to the laser-cutting machine. After the triangles are produced (four hours of laser cutting) all the parts are manually joined together with cable ties, (from one to one and a half days). The one hundred triangle stool is produced on demand using facilities at Metropolitan Works in London and each piece is unique. (Riccardo Bovo, 2012)

Lynne MacLachlan, Greg Saul/Tiago Rorke team and Peter Musson also use scripting and processing as a design tool. Peter Musson's hammers sit next to his home-built CNC milling machine and his computer is a tool to understand and develop the form that fits his clear image of the finish piece, which comes from his understanding of applied art and material. He is having a lot of fun playing with the visual programming language of Grasshopper to develop and produce forms, allowing him to do more than imagine or try to get them right with hours of modelling. He does still see himself as a craftsman with craftsmanship ideals. However, he believes that skills are not lost, they just shift and focus on what is important to the present, and that knowledge and information are transferable across disciplines.

'ANTLER' CHAIR PROTOTYPE V3 BY GREG SAUL AND TIAGO RORKE. Created using their SketchChair App. 12mm Baltic birch plywood. 478 x 815 x 590 mm. Diatom Studio, London, UK. 2011. Photographer: Greg Saul.

SKETCHCHAIR

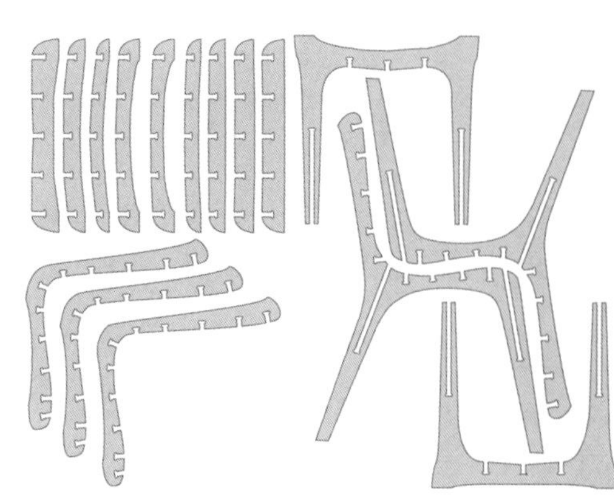

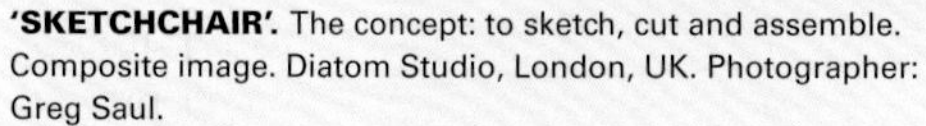

'SKETCHCHAIR'. The concept: to sketch, cut and assemble. Composite image. Diatom Studio, London, UK. Photographer: Greg Saul.

SKETCHCHAIR CUTTING PROFILE for uploading to an online digital manufacturing service

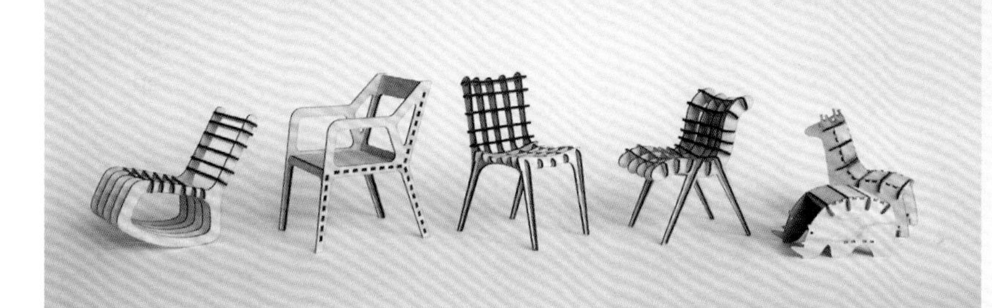

THE CHOSEN SERIES OF DESIGNS FOR THE KICKSTARTER CAMPAIGN PLEDGE REWARDS. From left: Rocking Lounger (co-designed with Nadeem Haidary), Edge Chair v2, Antler Chair Miniature v5, Stroke Chair, Twiggy and Rocksteady; scale 1:9. 1.5 mm Finnish birch plywood. Diatom Studio, London, UK. 2011, Photographer: Tiago Rorke.

SketchChair (www.sketchchair.cc) by Greg Saul and Tiago Rorke, was born out of a passion for tinkering, a desire to create accessible digital tools that anybody can use to design and make their own chair using the wealth of technologies becoming available and accessible. SketchChair is an international collaboration, built with the help of various other open-source projects (including the Processing programming language and the J. Bullet physics library) so that physical SketchChairs can be made from automatically-generated cutting profiles. The project was successfully crowd-funded on Kickstarter and released as a free and open-source system. SketchChair has a simple 2D drawing interface allowing anyone to easily design their own digitally fabricated furniture, to simulate sitting on it, customise it for comfort and automatically generate the structure and test its stability. Designs are shared in the online gallery and any design can be downloaded and edited, to evolve, refine and modify. So designs take on a life of their own as they pass through the hands of various authors and fabricators. The final SketchChair designs are uploaded to an online digital manufacturing service (ponoko.com), or sent to a FabLab/ local community workshop for parts to be cut using CNC routers, laser cutters or paper cutters. The parts are easily assembled by hand.

Greg started SketchChair in collaboration with the JST ERATO Igarashi Design UI Project, a research group in Tokyo, Japan, where interactive systems are being developed to cultivate end-user creativity and self-expression. Their approach to product design is like open-source software and creating designs is important in the process of developing software.

5 Applying 3D digital technologies

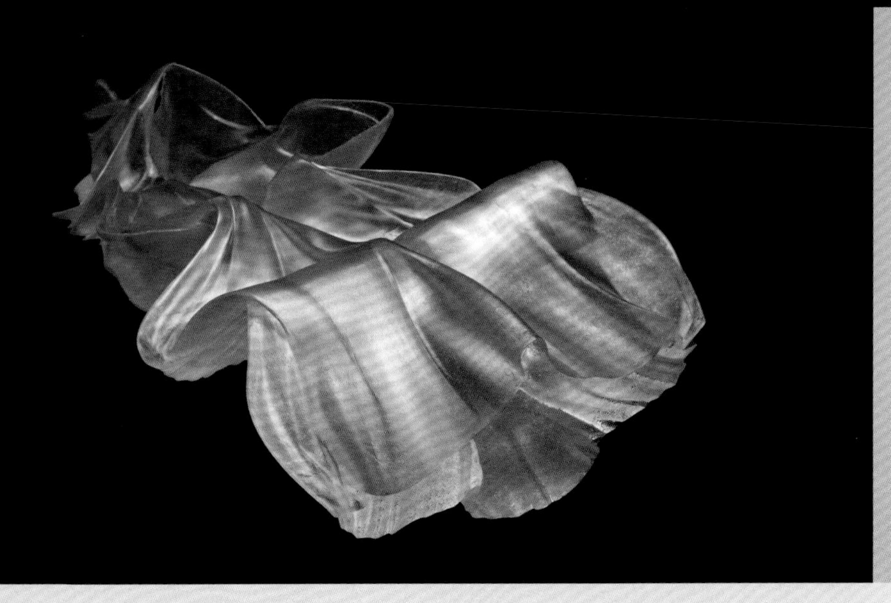

'FLIGHT' BY GEOFFREY MANN. Form and motion captured and translated into 3D data and 3D printed®. Photographer: Sylvain Deleu.

PART 1

3D digital designing

The computer is just a tool; it doesn't create innovation...it's how you apply it...it's a facilitator for where I want to go ... the technology should not overcome the final piece... — Geoffrey Mann

Geoffrey Mann's birds and moths are quality — Zac Eastwood-Bloom

I distinguish two types of 3D-design program: 3D CAD (computer-aided design) developed to design products and mechanical parts for industrial manufacture, and buildings and molecules, and 3D modelling software, which has fewer constraints. The difference between these programs is blurring as early 3D modelling tools merge with more prescriptive engineering-based CAD tools.

Objet RHINO GAYLE MATTHIAS Dr Cathy Treadaway Geoffrey Mann SPATIAL IMAGINATION commitment INTERFACE 3D machining GENERATIVE DESIGN 3D geometry CONSTRAINTS Photogrammetry SolidWorks CLAYTOOLS Algorithms LASER BEAM

TSplines
SENSABLE
MATERIALISE
GEOMETRIC MODELLING
DATA CAPTURE
reverse engineering
PARAMETRIC MODELLING

3D DIGITAL DESIGNING Johannes Tsopanides controls
INTELLIGENCE OF MAKING Monika Auch
DIGITAL FORM STL AUTONOMATIC
Stephen Bottomley DIGITAL
JOHANNA SPATH Volumetric modelling
Grossman TECHNOLOGY
3D printing INSPIRES AND
SSLE Zbrush MOTIVATES
CAD/CAM TAVS JORGENSEN
Wireframe JUST A TOOL
TRIANGLES MARLOES TEN BHÖMER

POINTS
OTHER SYSTEMS FOR CREATING 3D OBJECTS
3D MODELLING
SURFACE MODELLING
GRASSHOPPER
SOLID MODELLING

3D DIGITAL FORMATS MAGICS
FOR 3D OBJECTS Freeform
HYBRID SYSTEMS
3D SCANNING
123DCapture
LINES MeshLab
Polygon
CURVES

CAD and 3D modelling

Learning CAD can be a big commitment, 3D modelling less so. A program such as SolidWorks takes 2,000+ hours to reach professional level – hard to justify as a designer maker. Most of the work in this book is CAD designed – designers tend to stick to the package they used at college, classes/workshops or in work when they first started because there were few alternatives. Also they had technical assistance and the program offered was probably within the context of their subject, be it architecture, product, jewellery or textile design. It might not always be the most suitable package but it is hard to switch when so much is already invested.

It is now relatively easy to get into 3D design and 3D modelling – not only are there more software packages available than previously, there is also more variety, good functionality at lower cost and more ways to learn. There is also better access to 3D digital production such as computer-controlled machining and 3D printing to get models made.

Anyone seriously going for CAD commitment will already be familiar with this way of working, with CAD aesthetics and what different packages offer, gauged from colleagues, videos, online galleries of users' work, forums and blogs – sufficient to understand what you can do with it. Blogs such as Fabbaloo and 3D Printer and the websites of distributors and service companies providing milling, 3D printing, scanning and casting are great up-to-date resources. They offer useful observations regarding models and transfer files for 3D machining and 3D printing, listing software (note, for instance, that Blender has a steep learning curve yet is also the most widely used open-source 3D modelling tool), input devices (2D and 3D mice, tablets and styluses, haptic devices, gesture capture, holography and so on.

Generally, the more functions in a package, the more complex the UI (User Interface with multiple drop-down menus, symbols and icons) and the

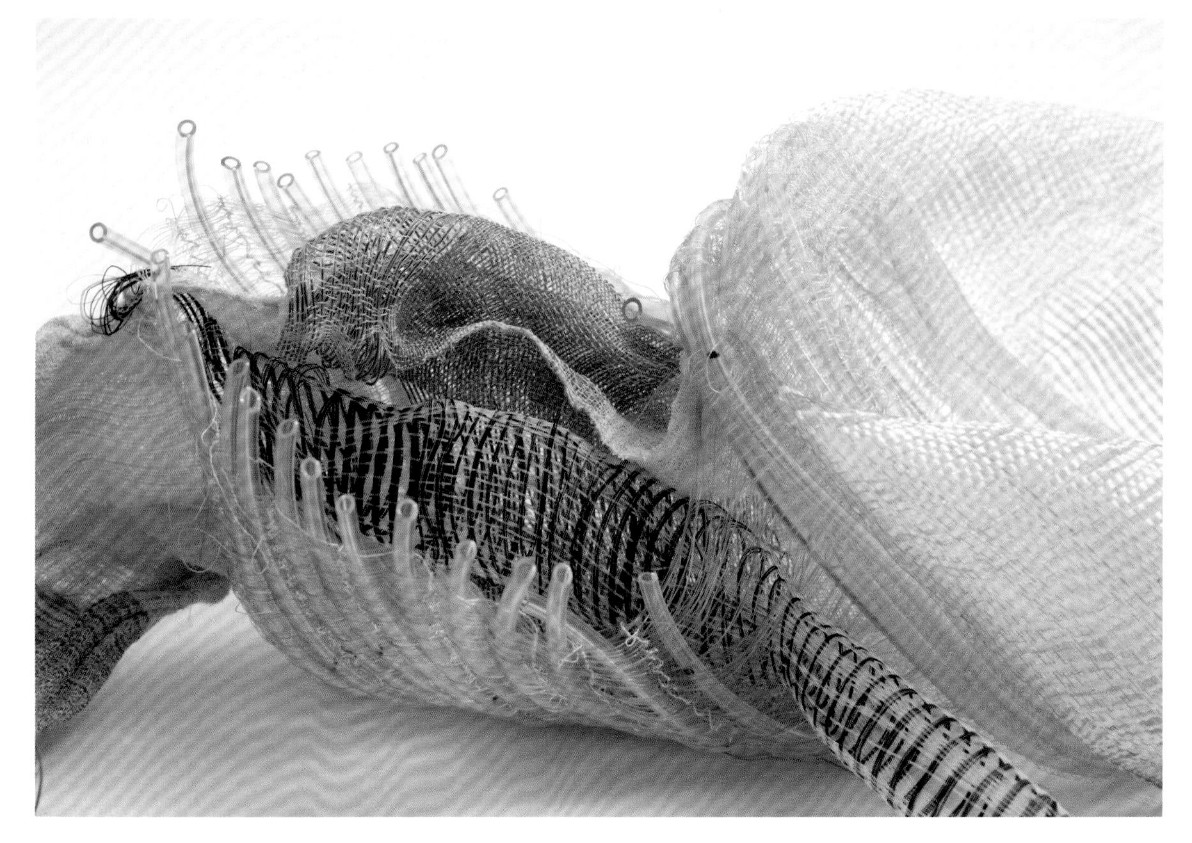

'NEUROTUBE' (DETAIL) BY MONIKA AUCH, from series Proxyclones material – plastic yarn, paper yarn, tubes, horsehair. 45 x 30 x 40 cm. 2009. Photographer: Ilse Schrama.

steeper the learning curve. Customising the UI can help by displaying and re-positioning the most used menus and setting up hot keys to suit. Many shortcut key functions (Ctrl+C/Ctrl+V for copy/paste, Ctrl+A for select all, Ctrl+X for delete, Ctrl Z for undo, etc.) are already pretty standard.

Applying CAD and 3D modelling

Geoff Mann is an artist, designer and lecturer and with his knowledge of materials, processes and craftsmanship he has all the terminology needed to collaborate with skilled technical practitioners and closely direct the crafting of his work, challenging and pushing boundaries to realise his artistic visions in glass, metal and ceramics.

CAD is a fundamental tool for Geoff. He has built up sufficient working knowledge of different types of program, for example, cinematic stop-motion techniques, learning only those functions he needs in order to visualise his ideas and to realise his work. He rarely sketches, going rapidly from exploring the concept in words, idioms and photographs, and rough models in ceramic to working on screen. Rhino as a CAD package provides him with most of the functions he needs to realise his ideas. As he says in an interview with Cathy Treadaway: 'I like Rhino because it lets me make mistakes'. (Cathy Treadaway: Studio visit – StudioMrmann, interview with Geoffrey Mann, Edinburgh, 14th January 2011).

CAD/CAM processes and products are considered devoid of feeling. Monika Auch, a textile artist, described this to me. To Monika, whose former profession as medical doctor keeps inspiring her, it seems that either you are a 'traditional' craftsperson and thus against CAD or you are in favour – experimental, fast and dynamic. Monika has found an in-between position, a fusion of both approaches.

Because time is a treasured factor to Monika, she realised early on that CAD/CAM weaving promised the end of frustration and feeling hemmed in by weaving's slow process, yet after many metres from a Dornier machine the novelty faded. It was still '2D' fabric, albeit beautiful and unusual.

Monika is fascinated with 3D embryologic growth: the possibilities of 3D woven shapes and technological developments. Her 'eureka' moment about employing weaving software for intricate, 3D-layered weaves happened while she was taking a CAD/CAM weaving course in 1992. Having studied multilayered weaves, it has taken her several years to figure out how to use the program and subvert the software from 2D shapes and patterns to layer functions to simulate the weaving process through as many as 12 different layers. She plans and then concentrates fully on the work that her hands do to build a form, combining three elements:

- good spatial imagination: mentally transcribing, translating and planning 3D into 2D on the screen and the moves in weaving;
- computer-aided designing acting as a guide;
- intricate tacit knowledge, tactile sense and excellent dexterity acquired by touching, 'knowing' and valuing the behaviour, properties and emotional meaning of many materials in (weaving) constructions, pushing boundaries by putting them to unusual tasks.

Her research is about the intelligence of making, the 'intelligence' in our fingertips and our hands, which sometimes work faster than our minds, the resulting work often looking different from what we imagined or planned. For Monika, that means her final 3D shapes only emerge when she releases the woven form from the loom, because materials will always respond differently, dependent on interacting factors. It simply isn't possible to control or predict this process.

Monika's earliest memories are of the thrill of making things with her hands, playing with and experimenting with materials. This sensitivity, technical skill and dexterity acquired over years is, she feels, being lost in our increasingly visually-centred world. Even with the help of computers we need the necessary material sensitivity to be able to create. However, she does wonder whether maybe humans will acquire new neurological pathways instead. (Edited version of Monika Auch's original statement in 2010, revised in 2012.)

Types of 3D formats for creating 3D objects

3D CAD and 3D modelling software packages use different modelling architecture such as wireframe, surface and solid modelling, and these determine what the software is capable of providing and therefore what it is used for.

Wireframe-type objects are solid objects consisting only of lines, points and curves. As the least complex type for representing and creating 3D objects, wireframe is used for 3D constructions where a high screen frame rate is needed, e.g. to rotate objects or for real-time simulation and animation. This format is also widely used in programming tool paths for computer-aided manufacturing.

Surface modelling: wireframe is not to be confused with polygon mesh, which is used for 'surface modelling' and made up of triangles, quadrilaterals and other polygons. This mesh defines the object's form and for CNC milling and 3D printing this mesh model must be intact, i.e. 'solid'. That means no holes, no floating unattached polygons, and all edges and faces 'continuously connected' to each other, edge to edge to edge to edge all the way around and over.

Solid modelling is classic CAD for designing products for manufacture as an object's 3D geometry as 'a consistent set of principles for mathematical and computer modelling' (http://en.wikipedia.org/wiki/Solid_modeling) is fully described: it can be rotated and viewed from all

RYGO PENDANT
BY BATHSHEBA GROSSMAN.
Modelled mathematically from a gyroid (a triply periodic minimal surface) using Mathematica and Kenneth Brakke's Surface Evolver, clipped with an ellipsoid and perforated. 3D printed by i.materialise in alumide (part of .MGX 2011 Summer Collection). 39 x 28 mm. Photographer: B Rodnes.

angles, manipulated, exploded, sliced to reveal internal structures and parts, and so on. Solid modelling includes more and more functions, for simulation, planning, verification of processes needed for machining and assembly and for parametric modelling.

Parametric modelling is developed for specific usage as algorithms both control and prevent manipulations and constructions which would violate specific principles, rules and physics. An example is the parameters for architectural categories (such as doors and windows) which are given properties and physical constraints specific to their type to define their complex and essential relationships with other objects such that the door must be in a wall and not violate this principle by being placed in the floor.

Although modelling using geometry is core to CAD and CAM (computer-aided manufacturing), geometric modelling is widely used in applied technical fields (engineering, architecture, geology and medical image processing) using maths and algorithms to describe and generate two- and three-dimensional shapes.

Bathsheba Grossman, well known in both geek culture and in the 3D printing industry for her use of geometric modelling, is a jeweller and artist exploring the region between art and mathematics for creating small sculptures whose execution is born from an art aesthetic.

Although her sculptures are based on symmetry point groups, a very mathematical idea, Bathsheba doesn't use maths directly, preferring to draw and physically model in clay before moving into CAD to model the complex geometries. 'My work is about life in three dimensions: working with symmetry and balance, getting from the origin to infinity, and always finding beauty in geometry. Strangely, it works that way ... What is beautiful converges to what is mathematical.'

Her recent sculptural pieces use a more fluid method of modelling, working with TSplines for Rhino where algorithms keep the maths underground. She uses multiple applications for different tasks: TSplines for Rhino for symmetry and flow, Rhino for lofting lines and its mesh format to export into Zbrush, a mesh refiner, to enhance surfaces and texture them, Grasshopper to manipulate code directly, Magics from Materialise and MeshLab to fix models before sending out to be 3D printed. This involves a lot of what she calls 'circuitous workflow caused by trying to make incompatible software work together.'

Volumetric modelling uses voxels, which are cubic pixels, whose positions are inferred relative to other voxels. Voxel modelling is akin to using a malleable clay-like material that can be pushed around, bits can be added and subtracted. For ceramacists it is more suitable than mesh- and geometry-based types, particularly when used with a haptic device, e.g. Sensable's Freeform and Claytools.

Other systems for creating 3D objects

3D computer graphics (CG) is different from CAD and 3D modelling as many CG programmes use vector triangles to represent the part of an object that is seen. These vector/pixel entities are not encumbered by dense geometry so are very effective for simulating and faking 3D on 2D pixel-based screens: full-colour textured images are projected onto fast changing entities modelled in vector triangles and continually re-rendered, giving the illusion of solid objects. Illusions cannot be 3D printed and milled – they need to be converted into a solid model capable of being sliced into layers.

Although 2D is so different from 3D there are processes that can use 2D information to create 3D data. Towards this is one system that takes 2D images (Bitmap or JPEG) and converts them to greyscale. A 'heightfield' algorithm gives each pixel a scaled value for each level between white and black, from which a low-relief mesh (2½D) is constructed and given substance in a CAD package.

▣ KATHRYN HINTON: DIGITAL HAMMERING

Kathryn Hinton

Jeweller and silversmith Kathryn Hinton created 'digital hammering' as a more intuitive physical/digital 'silversmithing' tool, focused on merging the traditional techniques of hand forging and raising with CAD (Rhino), and 3D printing for casting.

KATHRYN HINTON'S DIGITAL HAMMERING SYSTEM FOR FORMING. 2010. Image © Kathryn Hinton.

DIGITAL HAMMERING: STERLING SILVER SMALL CAST BOWL 1. 3D printed in wax and cast with faceted effect, build lines from printing and texture from casting also left as a signature of process. Image © Kathryn Hinton 2011.

LARGER FACETTED PRESS-FORMED BRITANNIA SILVER DISH. Sheet metal formed on hydraulic press using male and female forma milled directly into tool steel from digital mesh. Image © Kathryn Hinton 2010.

'The Craft of Digital Tooling' was her MPhil research project and to this she applied her understanding of traditional methods of forming using steel or wooden blocking hammers and a sandbag to develop a unique digital hammer and software interface that captures information about the physical strikes of a sensory 'haptic' hammer. She feeds this into a 3D computer design programme. Being more direct and immediate, effectively designing with each strike, her system has a new physical layer of interaction between hand and materials, between two and three dimensions to create a tool to experiment, explore and design different forms by manipulating digital mesh within a CAD programme. Smaller models are 3D printed and cast but as digital mesh is not restricted to any one specific material or process of manufacture, this system can be used for press forming.

▣ UNFOLD: GESTURE MODELLING

Also working towards a more inclusive system is Unfold, a spatial design bureau founded by Claire Warnier and Dries Verbruggen. Together with interaction designer Tim Knapen, they developed L'Artisan Electronique, a system to model digital 'clay' and ceramic forms by capturing the gestures we would use to shape and manipulate clay on a potter's wheel. Applications like these apply what we know so that we come with experience and skills that enable us to straddle the analogue and the digital worlds: to digitally model what can be 3D printed.

UNFOLD'S L'ARTISAN ELECTRONIQUE INSTALLATION FOR VIRTUAL MODELLING USING LASER SCANNING OF HAND GESTURES. Commissioned by Z33. Demonstrated at OpenStructures Milan, 2011. Photographer: Peter Verbruggen.

◄ A MODEL BEING EXTRUDED IN CLAY ON UNFOLD'S CERAMICS 3D PRINTER. Photographer: Kristof Vrancken.

▲ A VIRTUAL MODEL BEING 'HAND' FORMED ON UNFOLD'S VIRTUAL WHEEL. Screen capture by Unfold.

Text on this and the opposite page is an edited version of Kathryn Hinton's statement, sent to the author for this book.

More recently 'scanning' type applications have developed that use multiple digital photos of stationary non-shiny objects, taken in the round under decent lighting. Clever algorithms morph the images into a 3D representation and make these into shells or solid models capable of being 3D printed. One free cloud-based application is AutoDesk's 123DCapture.

3D scanning

3D scanners scan across and around an object's full surface, collecting accurate digital points from which a 'point cloud' is created. From this a triangulated surface mesh is generated to construct a 3D model. Although undercuts are tricky, scanning is used to capture models of odd, natural and highly detailed forms that would be difficult and expensive to generate from scratch using CAD; for capturing very delicate objects (such as archaeological artefacts) and very large ones (like buildings) that cannot be handled; for converting handcrafted models into digital ones (termed 'reverse-engineering') and for scaling up and down or to re-model.

Low-cost scanners use 'off the shelf' red or green laser pointers to project a laser bar onto the object while a webcam captures images of the laser's path. Software stitches the parts together to create the 3D digital object.

There are more sophisticated methods for use when precision is important: laser light 'no touch' systems; hand-held digitising arms that record the position of selectively collected strategic touch points; fully automated robotic probes moving systematically over the surface of an object and photogrammetry which uses photographs and software to extrapolate through multi-point triangulation the three-dimensional aspect of the subject. Hybrid systems combine lasers and photography for realistically coloured virtual 3D models of objects.

'SINEW' BY GAYLE MATTHIAS, COLLABORATING WITH TAVS JORGENSEN (TJ). Kiln cast glass and found ceramics. Process used scanning, digitising and 3D printing. H 15, W 53, D 47 cm. Photographer: Simon Cook

Gayle Matthias, a practising glass artist, is currently working alongside Tavs Jorgensen, a freelance designer and research fellow at Autonomatic, a digital research group at University College Falmouth. Their collaborative research is concerned with Rapid Prototype moulding methods for glass casting and they have devised a way to cast directly into reinforced printed moulds. For 'Sinew' they used digital tools to scan ceramic surfaces (Roland Picza 3D laser scanner, Microscribe digitiser) to get digital information and, using Materialise's Magics and Rhino software, created a form that can be translated into glass using 3D printing, moulds and casting. Gayle's intention was to redefine and rearrange broken ceramic sanitary ware, using cast glass to extend and exaggerate anatomical references, and to attach and accurately unite two ceramic forms in new, and intentionally awkward compositions.

In 2010, designer Marloes ten Bhömer worked with 3D printer manufacturer, Objet, to apply 3D printing technology to her craft of creating women's shoes in contemporary designs that de-structure and reinvent using non-traditional technologies and materials to manufacture. She had a shoe last scanned using an automatically controlled probe. This mapped the surface around which to model perfect-fitting 3D-printed shoes in Rhino. Using a printer with multi-material capability the whole shoe was built in one go in successive layers in two different photopolymer materials, providing different properties to give her 'Rapidprototypedshoe' both flexible and rigid sections for comfort.

There is no assembly but the shoe can be taken apart and reassembled to replace worn parts and perhaps add a differently coloured or shaped part. And of course, because they use 3D scanning, the shoes always fit the wearer.

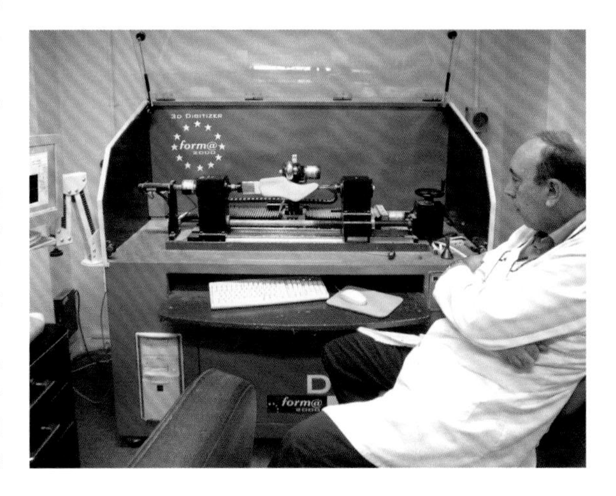

'RAPIDPROTOTYPEDSHOE' BY MARLOES TEN BHÖMER: a shoe last being scanned automatically for 3D digital form on which to model and her 'rapidprototypedshoe'. 3D printed in two different photopolymer materials. 2010. Photographer: the artist.

'SPOONING' BY STEPHEN BOTTOMLEY.
Using scanning, CAD, 3D printing and casting to create a new 'object'. 2009. Photographer: the artist.

Contemporary jeweller Stephen Bottomley used scanning (reverse engineering) to transform ordinary stainless steel teaspoons and plastic blister packaging for an exhibition he curated, marking Woolworths' centenary and closure in 2008. Dr Peter Walter at UWE laser scanned the packaging with a handheld Z-Scanner. The new 'spoon' formed in Rhino 4 and Solidworks was prototyped with Objet's Eden 350v 3DPrinted before sending the file to be 3D printed in resin, cast from this in precious metal, finished, its title 'Spooning' laser etched, and hallmarked 2009 Edinburgh. A tale of rags to riches. ('Re-worthit!' exhibition, project originator and manager: Stephen Bottomley; catalogue, 2009)

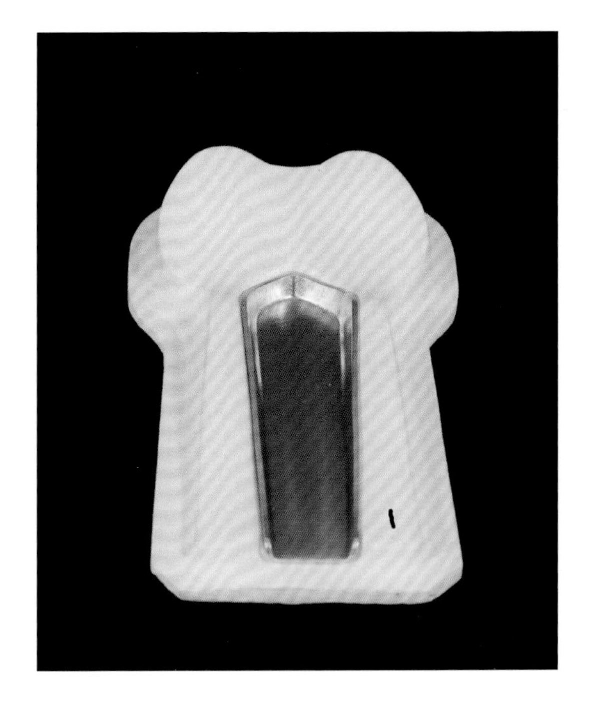

Using a Planar 3D scanner, Geoffrey Mann investigated the reflective properties of a Victorian metal candelabra. The laser beam is unable to distinguish between the surface and the reflection, so captured spikes represent the intensity of the metal's reflections. The resulting model for 'Shine' was 3D printed, cast in bronze, finished and silver-plated.

'SHINE' BY GEOFFREY MANN. Natural Occurrence series. Scanned, 3D printed, bronze cast (Powderhall Bronze), silver plated. 35 x 27 x 30 cm. Edition of 3 plus 1AP. 2010. Photographer: Nick Moss.

'DOGFIGHT' BY GEOFFREY MANN. Long Exposure series. Sub-surface engraving of three blocks of optical cast glass (75 x 10 x 35 cm installation). Vitrics, Paris. 2008 © photographer Sylvain Deleu.

Data capturing

In his 'Long Exposure' series Geoffrey Mann captures the motion and flight paths of moths and birds, making these usually invisible paths 'visible'. 'Attracted to Light' digitises the erratic flight of a moth around a light source which is 3D printed to create a hanging lamp. 'Dogfight' similarly captures the entwined interaction between two moths but uses sub-surface laser engraving (SSLE) to make this 'visible' within blocks of glass. With SSLE, a laser is focused within the blocks of high-grade optical quality crystal glass and fired to create a myriad of small fractures, the pattern controlled by 3D digital data. As the glass is perfectly clear the cloud of fractures appears suspended in 'air' like wisps of smoke.

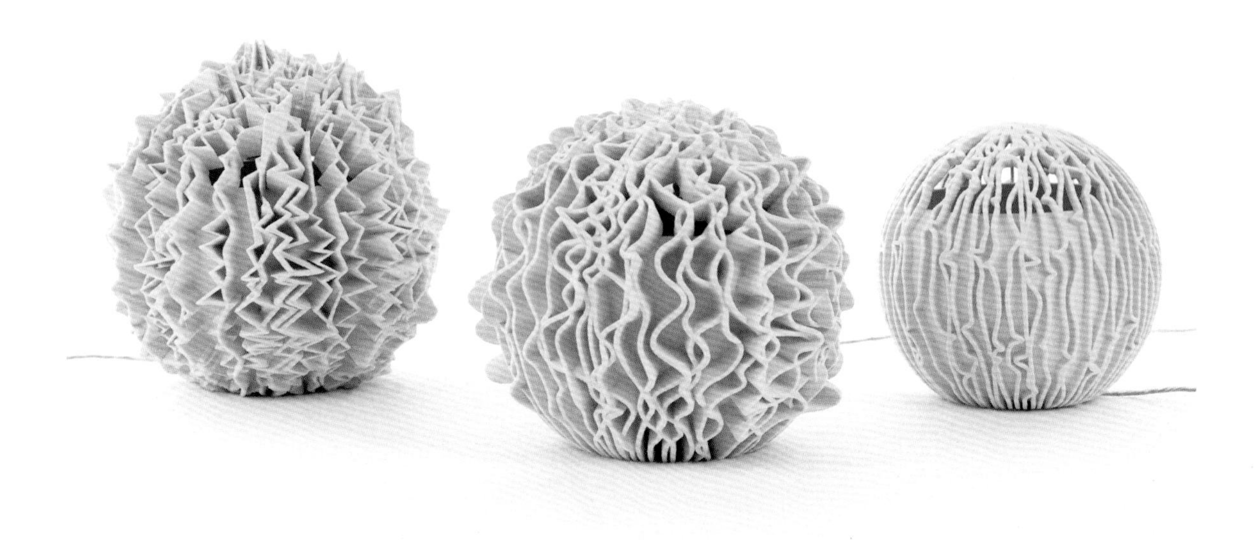

'CLOUDSPEAKER' BY SHAPES IN PLAY/JOHANNA SPATH, JOHANNES TSOPANIDES. Capturing and reflecting the music taste of its owner. 3D printed in SLS by EOS. (Disseny Hub Museum's permanent collection, Barcelona) 2008. Photographer: JohannesTsopanides.

Using PROCESSING (open source Java-based programming language) designers Johanna Spath and Johannes Tsopanides capture abstract information – behaviour, movement – from which forms are constituted, i.e. louder, sharper sounds reflected in bigger objects, angular forms, and 3D printed as individualised products. They founded their design studio, SHAPES iN PLAY, to focus on generative design which offers a new kind of access to explore the exciting borders where design meets technology, crafts and art. They are part of a community that is exploring what happens when designers leave space for interaction between user and object, finding new ways of designing and making to challenge boundaries between maker and consumer.

'SOUNDPLOTTER' BY SHAPES IN PLAY / JOHANNA SPATH, JOHANNES TSOPANIDES. Captured sound input progressively modifies a layer at a time visualising sound and time based personal and emotional meaning. 3D printed in SLS by 3DGiesserei Blöcher. (Disseny Hub Museum's permanent collection, Barcelona) 2008. Photographer: Johannes Tsopanides.

Data formats

To maintain the integrity of any model and transfer unambiguous data for 3D milling and printing, file types need to conform to industry standard file transfer formats which include IGES (Initial Graphics Exchange Specification), DXF (Drawing Exchange Format) and STL (Stereo lithography). STL is the most common which all CAD systems and most 3D modelling software are capable of producing and with most it is as simple as selecting 'File', 'Save As', and 'STL', so you need know no more.

3D models in formats such obj, .stp etc can also be exported for 3D printing and CAM, but as acceptance differs you should check this out. For example .3ds is the file extension for objects modelled in 3D Studio but not all service companies have the capability to translate this for printing, as converting to STL file can be complicated.

The number of triangles defines 'resolution': the more triangles, the smoother the model. The more complex the surface, the more triangles. Both create bigger files so more computation is required. As a guide, models in the range of 1MB to 5MB produce good parts.

Triangulated models can and do have unwanted and problematic 'holes', floating, unjoined parts and unattached edges which render the model incomplete and possibly not viable for 3D printing and CAM. Service companies, bureaux and high-end CAD packages have sophisticated algorithms to uncover problems and mark them for fixing and repair, and there is free software to mark uploaded STL files, even repairing minor problems. But taking care throughout the design to avoid problems and clean up as you go is still best. So with a makeable model, preferably in .stl format, what comes next?

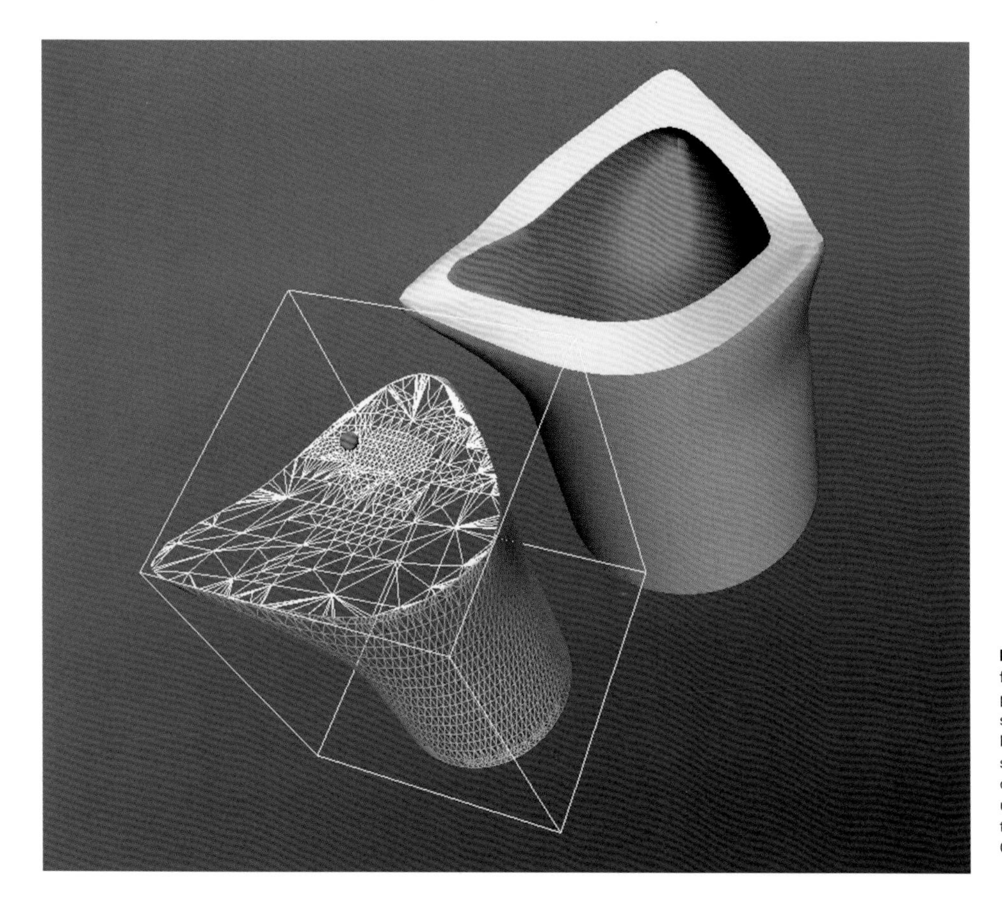

MESH MODELS from 3D modelling package, Cloud9, showing 'resolution' being enhanced by submeshing, i.e. dividing large triangles up to two, three, four times etc. Screen Capture: Anarkik3D.

PART 2

3D-Computer-aided production: machining and 3D printing

Steel
PAPER
titanium
GOLD
electroform
Gilbert Riedelbauch
x,y,z data 3D PRINTING
Esteban Schunemann
METHODS, MATERIALS AND APPLICATIONS
OPEN SOURCE
TINKERING
GORDON BURNETT
subverting
photosensitive resin
GRANULAR POWDER
LIZZIE ARMOUR
Adrian Bowyer
POLYAMIDE
Joris Peels
personal 3D printers
DANIEL WIDRIG
JAMIE MILAS
master pattern
Sarah Silve
i.materialise
extrusion
TITANIUM
Peter Weijmarshausen
UNFOLD
VIRTUAL 3D MODEL
Markus Kayser
CNC machining
Farah Bandookwala
QUICK AND DIRTY
Anthony Tammaro
TANGIBLE MODEL
TECHNOLOGIES
SUBTRACTIVE TECHNOLOGIES
3D COMPUTER
AIDED
PRODUCTION
DEFINITIONS AND TYPES
PHYSICAL MATERIAL PROPERTIES
Zac Eastwood-Bloom
PONOKO
ADDITIVE TECHNOLOGIES
LIONEL DEAN
Alissia Melka-Teichroew
COMPLEX FORMS fabbing
Customise Jae-won Yoon
silver
sintering
BRONZE
Shapeways
Sculpteo
CASTING
Bathsheba Grossman
Kraftwurx
PRICING OF MODELS
Peter Musson
Finishing
David Poston

With computer control, systems become very powerful tools for industry, but the designer maker can become removed from 'what is implied by craft production [...] an intimacy between producer/object, object/consumer, producer/consumer' (Johnson, 1997, p. 93)

The open-source ethos (that is, development through bartering and collaboration, with source-material, documentation and end-product available at no cost to the public) is very active in the area of 3D computer-aided production with growing pockets of pioneering independents and Fablab communities (fabrication laboratory) actively returning to a hands-on spirit – they are improving and opening up access to CNC machining and 3D printing. Not only are DIY milling machines and Personal3DPrinters (P3DP) developing rapidly with faster speeds, higher resolutions for better surface finish and expanding range of materials, designer makers are setting up shop with their own personal kit.

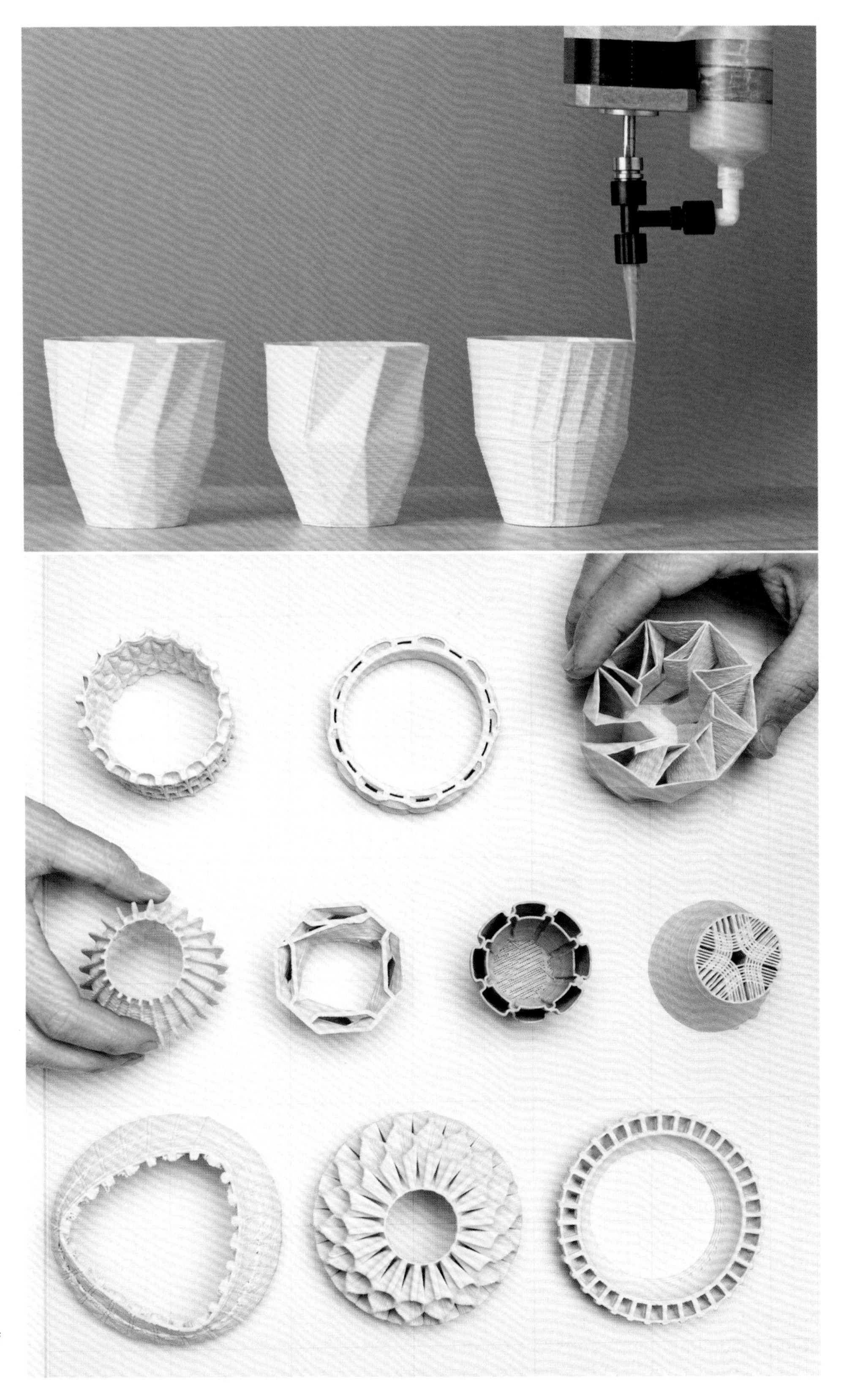

UNFOLD'S 3D PRINTER for clay: based on the open source RepRap project, a printer kit from Bits from Bytes was modified to extrude clay. 2012. Photographer: Kristof Vrancken.

3D PRINTED JEWELLERY PROTOTYPES BY HOT POP FACTORY. Designed using Rhino/Grasshopper and made in ABS Plastic on Hot Pop Factory's own Makerbot Replicator 3D printer. 2012. Photographer: Hot Pop Factory.

Peter Musson's metalsmithing workshop is his laboratory where he develops and adapts his own CNC system for his needs, for fine scanning as well as 3D printing in metals. He describes this as,

a good way to waste time, what with playing with it! Who said CNC would save time in the crafts! Ha! I feel CNC system is very important to my work. One: it is cheap to run, two: it gives me freedom and three I have a higher feeling of correctness.

This is just one exciting part of the larger picture that is computer-aided production which, as Nidhi Shah of Imaginarium puts it, is about:

automatically producing finished products by using computer-controlled production machines ... CAD and CAM work together in that the digital 3D model generated in CAD is fed as input to the CAM machine.

Definitions and types of computer-aided production

There are two types of 3D computer-aided production: 'subtractive' systems (machining to remove material from a solid block) and 'additive' systems (3D printing that selectively constructs an object by adding material). Both operate in three axes: X = left to right, Y = back to front and Z = up and down.

'Subtractive' is familiar: lathe-turning, milling and cutting machines remove material like wood, metal and composites (e.g. plaster and polymers) from a block. For eons it has been hands-on and direct so designer makers already have vast knowledge of physical material properties and an understanding and feel for the idiosyncrasies and characteristics of different ones, such as hardness, growth patterns, pliability, grain, stickiness. McCullough describes working this way as 'reacting to a dense continuum of possibilities', since:

every material has tolerances, within which it is workable and outside of which it breaks down... An experienced craftsman knows how to choose the right medium and to push it as far as it will go – and no further. (McCullough, 1996, p. 198)

Additive manufacturing (3D printing) is relatively new and is mainly used in industry as a prototyping process. Its usage has extended to new users in universities, workshops, schools, and now to anybody and everybody as consumer services online and with the introduction of Personal3DPrinters into the home. In 2010 Joris Peels said:

Thousands of people have bought and made RepRaps and Makerbots and tens of thousands use 3D printing services such as i.materialise to 3D print their own creations. 3D printing is appearing on TV, in newspapers and magazines for the first time. Millions of people have learned about the technology this year as media coverage and demonstrations at fairs have reached them. This is the 'year of acquaintance', where the world finally meets this 25-year-old technology.

Being new, additive manufacturing needs to be explained. The principle is straightforward, based on the concept of 'bacon-slicing' a virtual 3D model (1 in the illustration below). These 'slices' are sent as x,y data (2) to the 3D printer which 'prints' out these slices precisely (3a, 3b) layer by layer by layer, one on top of the previous one so the original sliced virtual object is replicated, constructed vertically (z axis) as a tangible model.

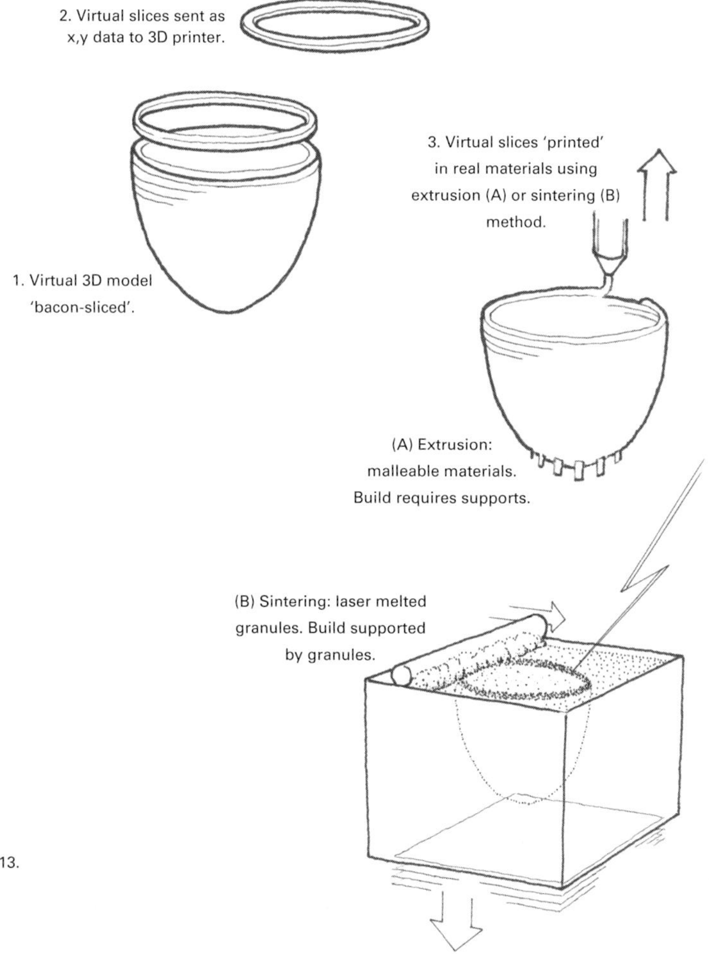

THE CONCEPT OF 3D PRINTING
© Bjorn Rodnes. 2013.

Different systems use different materials and different methods of constructing in layers, for example, extrusion (3a) and sintering (3b), (see pages 110–13), providing different resolutions/ degrees of finish and size of build. Each system has strengths and weaknesses, likewise the objects produced.

In an interview with Josh Wolfe for Forbes, Peter Weijmarshausen, CEO of Shapeways said that

3D printing allows you to make incredibly complex designs at no additional cost: interlocking components, naturally hinged parts, semi-translucent surfaces, and even objects that can move...

New methods and hybrid machines are being developed: ones that build more organically – not in prescribed layers but with each unit of material (brick scale or at nano level) deposited accurately by nimble robotic fingers, ones that combine additive and subtractive methods – waterjet cutting and adding material by welding parts or by fusing powder, all at high-speed. Meanwhile we have the clear-cut distinction between two systems.

Subtractive technologies

There are three main types of machining:

- Three axes: straight, no undercutting – the machine bed and model move in two axes (x and y), the cutting tool moves in z-axis.
- Four axes: similar except the model is rotated in one axis and the cutter moves in the x-, y- and z-axes. Rounded forms are possible (with minor constraints).
- Five axes (similar to four axes except the model is rotated in two axes and the router, or waterjet, moves and/or rotates in the x, y, z axes. Fully rounded forms with complex profiling and undercuts are standard).

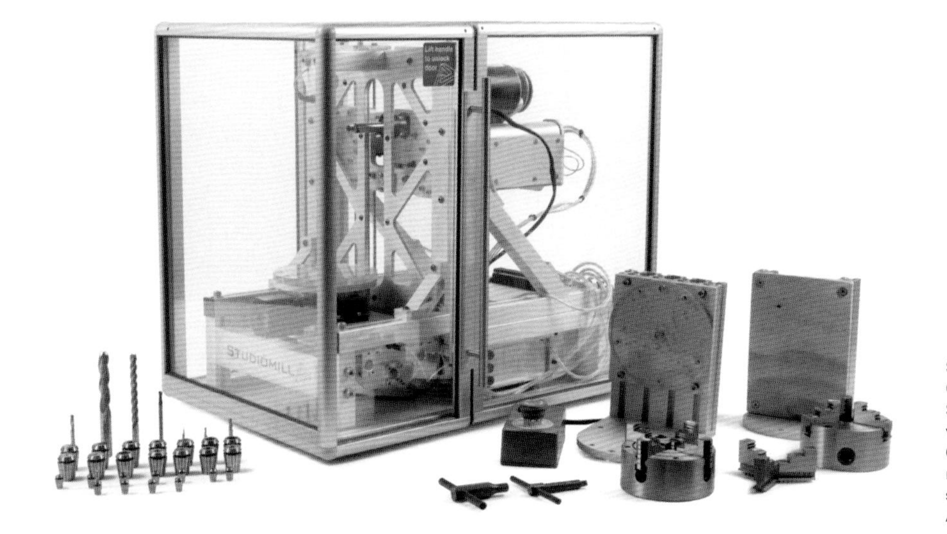

STUDIOMILL FROM CORE PD GROUP. Studiomill D160 – versatile multiaxis CNC, perfect for realising complex shapes. Photographer: Alex Dobbie.

'BRAZIL': LIMITED EDITION ARMCHAIR SERIES BY DANIEL WIDRIG. Plywood, CNC machined. 65 x 90 x 70 cm. 2009/2010. Photographer: the artist.

Computer-aided machining (CAM) has become a very competitive process with great potential in the hands of designer makers for bespoke work, for rapid prototyping, and for making master patterns. Compact, affordable machines with rotary axis modelling capabilities and with a 'virtual' fifth axis (using specially developed software) are suitable for Fablabs and micro-businesses to make models and jewellery, cutting soft materials such as wax for lost-wax casting.

The skill is in programming the switches between cutting tools needed for different types of carving, milling and drilling, between roughing out the shape, refining and detailing it, and finally finishing it to a smooth surface. With softer materials like wood, a waterjet head can move in three axes and rotate in up to three to carve out the most intricate and complex forms. With hard materials such as metals, torque to the machine joints limits the rotational movement of the cutting head. Seeing some of these machines in action on YouTube and at industrial trade fairs is like watching a robotic ballet.

Daniel Widrig designed his Brazil chair in digitally nested parts to save on material. Once machined from laminated wood using a five-axis CNC router, parts fitted together, glued and hand finished, the whole form becomes one, 'flowing' up through its three legs into armrests and a back with surface patterns that emphasise the chair's layered structure and construction.

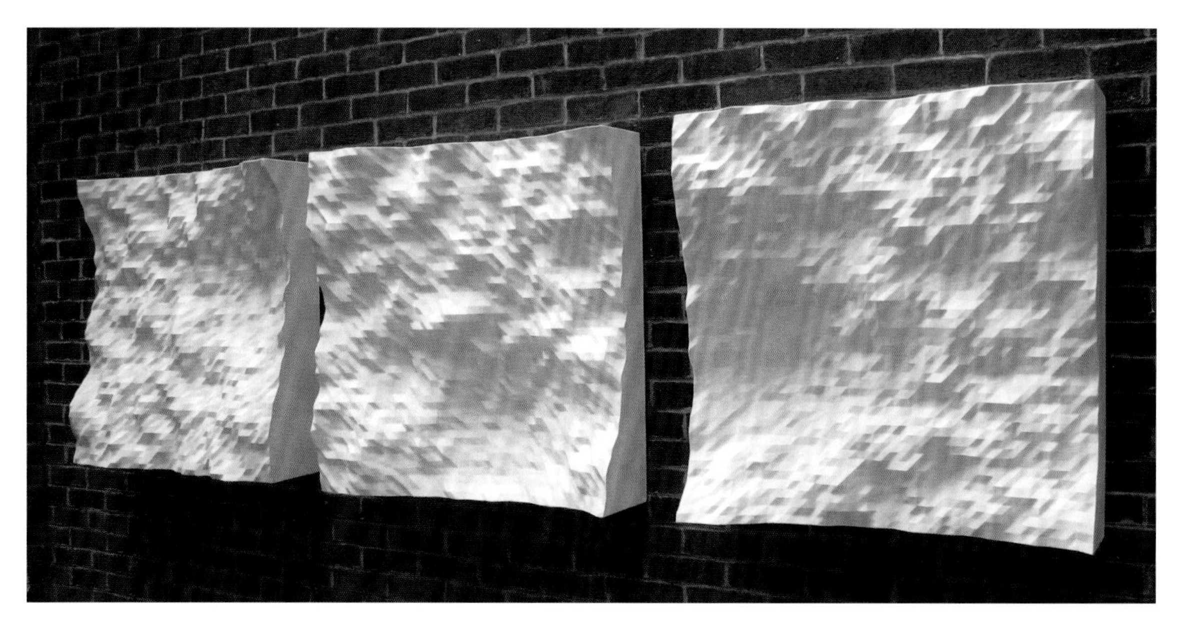

21ST-CENTURY LANDSCAPE TRIPTYCH, EXPLORING LOCALITY OF THE DIGITAL, BY ZAC EASTWOOD-BLOOM. Acrylic polymer and glass fibre. 1 x 3 m. 2010. Photography: Dominic Tschudin.

'INFORMATION ATE MY TABLE' BY ZACHARY EASTWOOD-BLOOM – exploring what happens when 'The Digital' crosses over into the material world. CNC milled beech. 37 x 96 x 59 cm. 2010. Photographer: the artist.

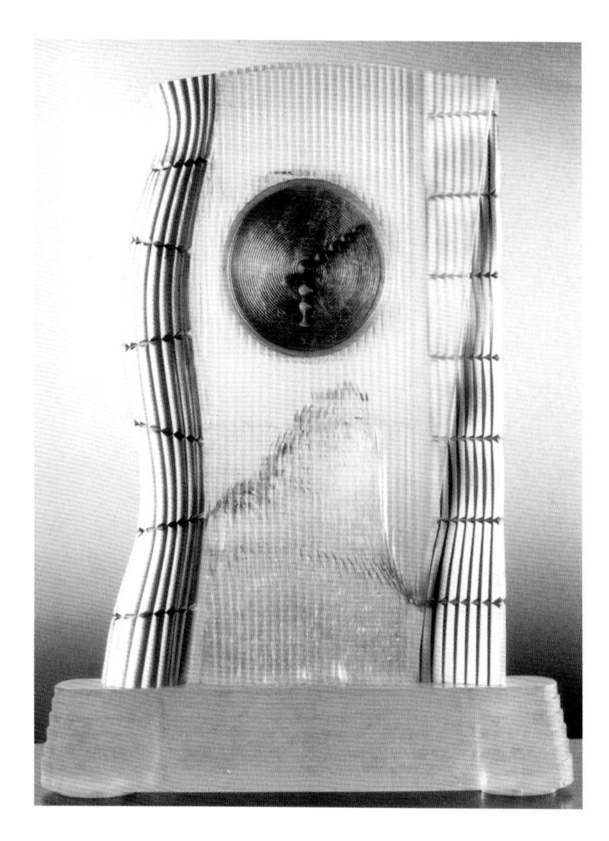

MILLENNIUM CLOCK FOR RESIDENCE OF SCOTLAND'S FIRST MINISTER BY GORDON BURNETT
Machined, anodised aluminium, 3D printing, and casting. 55 x 37.5 x 0.1 cm. 2001. Owner: The incorporation of Goldsmiths of the City of Edinburgh. Photographer: Shannon Tofts.

The syntax of process for making objects fascinates Zac Eastwood-Bloom: remove one stage and the final outcome is likely to fail. The Royal College of Art gave him means to investigate and connect digital making with historic processes of making such as ceramics, bronze casting and woodworking. He linked programs such as Terragen, Rhino, Illustrator, Maxwell and plotting software to explore digital geographies, developing surfaces into forms for 3D printing or CNC milling. He says,

My works examine the increasingly flimsy space between the real and the digital, offering a view of the sublime and the borderlines between the human and the digital whilst using contemporary tools and processes. Twenty years ago 'becoming one with the machine' meant a physical oneness with tools, yet in the 21st century this is becoming an increasingly psychological process.

For 'Triptych', his digital file was scaled to a metre squared and CNC milled into styrofoam from which a plaster model was made and the final work cast in a plaster polymer with glass fibre. Exhibited outside, the facets created subtly changing shadows throughout the day.

The preparation and lamination of the raw sawn planks of beech wood for 'Information Ate My Table' ran in parallel with the preparation of the digital model for production using milling. With the final finishing, sanding and lacquering of the beech there was a sense of poetics to the making process; a balance of digital and material processes.

Gordon Burnett is a proficient user of digital technologies and used a medley of additive and subtractive systems for a Millennium Clock commission in 2000. In context with his on-going search for unity and harmony between concept, the body, and materials and tools, the whole design evoked the tension between natural and man-made worlds. Handmade models tested scale and context before using 'Solidworks' to create digital drawings of the organic and geometric columns, 3D-printed in wax and cast in silver, and for CNC-machined wax for the clock centre and base, cast in crystal and resin.

3D PRINTING METHODS

NOZZLE SLICES HOT-MELT cut sheet OVERHANG FDM UV light INKJET SOLIDIFY MCOR S. Scott Crump Dr Carl Deckard harden GELS
MICRO-DROPLET 3D PRINTING SUPPORT STRUCTURES
Electron Beam Melting METHODS selective laser melting
DMLS DIRECT METAL LASER SINTERING SLICE PROFILES Binding granular SLM
Arcam AB SINTERING GRANULAR materials BUILD TRAY Professor Neri Oxman
DIGITAL 3D MODEL Objet Geometries MATERIALS STEREOLITHOGRAPHY THIN LAYERS
EXTRUDING FLUID MATERIALS MATERIALS FLUID DEPOSITING FILAMENTS 3DP
Chuck W. Hull low-cost photopolymerisation build chamber SLS
SLA 3D Printing MULTI-JET

There are five basic types of 3D printing:

1. Depositing/extruding molten or fluid materials.
2. Sintering granular materials.
3. Binding granular materials.
4. Photopolymerisation (setting light-sensitive resin or gel).
5. Layering cut sheet.

For each of these types the process is the same: the slices of the digital 3D model hold the parameters which control where and how the layers of material are deposited to construct the tangible model.

On Hot Pop Factory's Makerbot Replicator 3D printer their Peaque Necklace is 'extruded' in ABS plastic. This takes about an hour, with every overhang delicately supported by stilts – some almost melting away before the print is complete. 2012. Photographer: Hot Pop Factory.

Deposition/extrusion works on plastics and anything fluid enough to extrude and print

Fused Deposition Modelling (FDM): an 'extrusion' process that uses very thin thermo-plastic filaments of materials (wax, polycarbonate, ABS plastic), deposited as a molten stream through a nozzle onto a base, then layer upon layer, builds with either the nozzle head moving up a notch or the build platform moving down. A second nozzle deposits a similar thread to build any structures needed to support overhanging parts of the model. Support material can be in different colours to make separation easier, dissolvable (in a soda bath) or easy to break off. FDM is used to create functional parts of any geometry: models in ABS are physically robust, and, for the jewellery industry, wax is used for the standard production method of lost-wax casting. FDM was developed by S. Scott Crump (Stratasys co-founder) in the late 1980s and commercialised in 1990. Most DIY 3D printers use the FDM technology as one patent has expired.

Other extrusion techniques: multi-jet modelling, micro-droplet fabrication (MDF). Wax or hot-melt materials are squirted or deposited as small drops from one, two or multiple nozzles in layers, building the part in solid material and the supports in a matrix that can be broken away easily and cleanly.

Sintering works with powdered material which can be spread and sintered (polymers, polycarbonates, metals, ceramics, starch, etc.)

▼ **DIRECT METAL LASER SINTERING: TITANIUM RING** DESIGNED BY ANN MARIE SHILLITO, 3D printed by i.materialise: left image is as it comes part cleaned, and right, after finishing (shown with gold/ diamond hand-wrought engagement ring by jeweller Teena Ramsay). 2011. Photographer: the author.

▲ **SELECTIVE LASER SINTERING: RING** BY BIRGIT LAKEN in alumide, by Shapeways. 40 x 37 x 10 mm. 2012. Photographer: the artist.

▲ **SELECTIVE LASER MELTING: BRONZE RING** BY ANN MARIE SHILLITO, 3D printed by ES Technology using CONCEPT Lasers machine. Ring cleaned up and partially finished. 2012. Photographer: the author.

In **selective laser sintering** (SLS) of powdered media granules are coated with a thin film of polymer binder and spread thinly over a moveable floor in the 'build chamber' where a laser moves across fusing/melting powder in the layer's profile and to the previous layer. The surrounding loose powder is compacted to support the part and any overhangs as the model is built up. The build chamber's floor then lowers a very small distance, new powder is spread over the area and the sequence is repeated until the final top layer is constructed. Excess un-bonded powder is vacuumed off and the 'green' 3D printed part is carefully extracted from the tub for further cleaning and infiltrating with liquid photopolymer and cured in a drying oven to strengthen and stabilise the piece to make it more robust. Sintered parts have a porous surface that can be smoothed, dyed or painted.

SLS systems have a greater range of materials as so many can be granulated and laser sintered together. Adding ground glass to the nylon powder gives it even greater strength for performance and durability. Other materials include ceramic, glass, zircon and silica sand for direct mould-making and materials are constantly being refined and new ones added to meet the requirements of industry. SLS was developed and patented by Dr Carl Deckard at the University of Texas in the mid-1980s under sponsorship of DARPA.

With **direct metal laser sintering** (DMLS), the process is lengthy and involves a lot of machinery and labour. For example, a 3D printing steel model is fragile and porous, so before the supporting loose powder is removed it is all put in a curing oven for 24 hours. Bronze powder is added and the model design goes into another oven for 24 hours where the bronze infuses into the design and hardens on cooling. Any supports are then removed and the model is given its finish.

In **Selective Laser Melting** (SLM), a laser beam is used to fully melt powder particles of stainless steel, titanium and other 'difficult to work with' materials. CONCEPT Laser GmbH invented and registered the process as lasercusing®, which was primarily developed to make moulds. Gold powder is fused in a nitrogen-filled build chamber and although the layering process is slow the parts are very dense.

In **electron beam melting** (EBM) Developed by Arcam AB, metal powder in a high vacuum is blasted with an electron beam to melt the powder, giving extremely strong fully dense parts free of any voids (distinguishing it from metal sintering techniques). Compared to a machined part EBM parts can be 60% lighter with less waste, which means lower environmental impact.

MINIATURE SCULPTURE BY ANN MARIE SHILLITO. Cloud9 model. 3D printed in photosensitive gel by Formero on an Objet printer for the 'Inside Out' Exhibition in UK and Australia. 6 x 6 x 6 cm 2010. Photographer: the author.

MINIATURE BRONZE CANDLESTICK BY ANN MARIE SHILLITO. 3D printed using selective laser melting, showing the supports partially removed, a relatively easy task leaving a good surface needing minimal finishing. 3D printed by ES Technology on CONCEPT Lasers machine. 2012. Photographer: the author.

Binding

'3D printing' (3DP) uses an 'inkjet printing' process with a print head moving about the build chamber, depositing a liquid that binds each thin layer of granules rolled on when the floor retreats one layer, repeating the process for the second and subsequent layers until the final top layer is produced. 3DP was developed at the Massachusetts Institute of Technology (MIT) in the late 1980s and licensed by several companies including ZCorporation. Zcorp's 2D colour inkjet methodology uses four coloured binders: cyan, magenta, yellow and clear, to mix and print colours onto the model's shell, for 24-bit, full-colour capabilities.

Photopolymerisation involves photosetting resins and other photosensitive materials and gels. **Stereo lithography apparatus** (SLA) is the oldest and most expensive 3D printing system, with best definition, developed by Chuck W. Hull (co-founder of 3D Systems) in 1986. SLA uses a laser to set layers of photo-sensitive resin. A laser traces a pattern on the thin layer of resin lying a on a raised platform inside the vat of liquid photo-sensitive 'resin'. Exposure to the UV laser light solidifies the profile. The platform moves down an incremental unit and resin floods the platform in a new fine layer. Loose and overhanging elements are secured by supports that are automatically created to prevent them drifting or breaking away.

When complete, the platform dramatically raises the object clear of the resin which drains out of any cavities. Finally the resin is cured to give a longer lasting prototype. The hardened resin is translucent so interior structure is easy to see. Parts created are very accurate, require minimal post-processing and are ideal for almost any prototype or master pattern for vacuum casting/casting because this process is more durable than powder and produces objects in high resolution to give very fine detailing and surface finish.

PolyJet inkjet technology also uses photo-sensitive materials. It works by jetting photopolymer materials in the slice profile as a thin layer onto a build tray, immediately sets it, and bonds it to previous layers by exposure to UV light. It was developed by Objet Geometries in early 2000. They are now leaders in multiple-material printing. In 2012 they launched a machine capable of 3D printing seven materials at the same time, each with different properties. Its potential was demonstrated in a collaboration with Neri Oxman, director of the Mediated Matter research group and assistant professor of media arts and sciences at the MIT Media Lab. She designed 18 amazing 'biologically' inspired 'Imaginary Beings: Mythologies of the Not Yet' which required advanced R&D from Objet to 3D printed in a mix of transparent, opaque and coloured materials. These were shown in the Multiversités Créatives exhibition, the Pompidou Centre, Paris in 2012.

Layering

Layering uses sheets of paper or plastic inserted one at a time into the build chamber to be cut by a metal blade. A binder is deposited to bond each cut profile to the previous layer. Models are finally broken out of the unbonded or sliced-up waste material. Mcor's paper printers are among the least expensive to operate. As the base material is abundant, models produced are cost effective and used to prototype large forms.

In 2012, Mcor launched their IRIS 3D printer producing low-cost, brilliantly coloured, photo-realistic 3D printed products and struck a deal with Staples Printing Systems Division to provide a new online 3D printing service called 'Staples Easy 3D'. (See the general links and websites on page 157 for more information).

Supports

All 3D printing systems must have some method to support models, particularly for the parts that overhang. Layering, 3DP, and sintering systems use the leftover material as support, whether sheets of paper or plastic, or powder. One advantage of these methods is that more objects can be stacked vertically as no supporting structures need to be built to hold them in place. The 'wow' effect is uncovering all the objects.

For sintering metal powders like steel, titanium and bronze, support structures are necessary as the compacted powder around the part is not sufficiently dense to support the heavier part and, with the immense amounts of heat generated, to keep it from deforming.

Stereolithography and fused deposition modelling (FDM) both require support structures. For FDM, two types of material might be deposited side by side, the support material being soluble/dissolvable for easy removal. Some systems automatically add supports and position models so the least number of supports are required. Contact points are minimal for easy and clean removal, essential for efficient production. Fine pointy-top rods snap off easily (just like Airfix) and a matrix is easily separated from the model. It is only with the smaller lo-tech kit-type Personal3DPrinters that general issues with supports constrain the design of objects.

3D printing technologies

In industry the term used is 'additive manufacture' (AM). Since 1980 other terms have been used including layered object manufacture (LOM) and rapid prototyping, which now covers all digital methods of prototyping. The term '3D printing' came into common use only recently.

Methods, materials and applications

Method and material are closely linked and determine the application used. Development of new 3D printable materials and better print quality is driving this technology and providing very exciting practical opportunities for designer makers.

Lionel Dean's 3D printed titanium 'Icon' pendants are the first products from automated production from his FutureFactories Company: an individualised edition of 100 one-off artefacts – each piece recognisably of a design type yet discernibly different through iterations. This 'mass produced' individualisation could only be accomplished with digital technologies and is the next new manufacturing system.

'ICON' PENDANTS BY LIONEL DEAN. 3D printed in titanium. 2008. Digital Image: Gareth Blakemore.

3D PRINTABLE MATERIALS

'FLICKER' INTERACTIVE CREATURE SERIES BY **FARAH BANDOOKWALA.** 3D-printed nylon, 3D-printed glass. Electronics, paint. 50 x 43 x 11 cm. 2011. Photographer: Tomas Rydin.

This sample list of 3D printable materials can by no means ever be comprehensive: development is just too rapid!

- photo-sensitive resins: translucent and clear
- plastics: ABS (what supermarket milk bottles are made from), polypropylene, clear acrylic, polyamide (nylon), glass filled polyamide, alumide, PLA (or polylactic acid, a bioplastics made from starch)
- metals: bronze, steel, including stainless, titanium, silver and gold
- composites: ceramic and glass
- clay, precious metal clay, rubber, plain paper, plain table sugar, wood fibre, wax, sand, foods, chocolate, human tissue, meat substitutes (lab-grown 'food grade animal protein')

3D printed precious metal clay (PMC): 2 tessellated rings – greenware (unfired PMC Pro) and fired to compare shrinkage on sintering. Research by Esteban Schunemann at Brunel University. 2012. Photographer: E. Schunemann.

Synthetic 3D printable materials have been developed for specific purposes such as:

- clean burn-out for 'lost-wax' type casting process
- flexibility, robustness
- dissolvable support material
- low cost for 'quick and dirty' prototyping
- extra-fine definition, for whiteness, for transparency
- qualities and properties essential for specific sectors:
  - hypo-allergenic materials and alloys for dentistry and prosthetics
  - withstand both the heat and pressure of vulcanising in rubber
- recycling: using materials that are already recycled, or biodegradable (bioplastics) and recycling on location

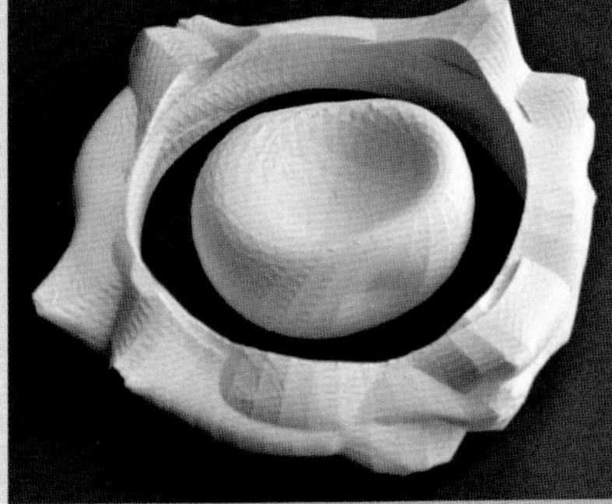

▲ **PAPER: BANGLE** BY **ANN MARIE SHILLITO** AND **'RED CORPUSCLE'** BY **SANDRA WILSON.** 3DPrinted in layers of white paper on Mcor's Matrix 300 3D printer. Photographer: the author.

▶ **PAPER: HECTOR SERRANO'S ALBERT CHAIR** 3DPrinted in layers of coloured paper on Mcor's Matrix 300 3D printer. Volume: 54 cc. Photographer: Cormac Hanley.

When Farah Bandookwala first started 3D printing, her peers assumed that she was 3D printing models from which to 'lost wax' cast. She says:

Given that selective laser sintering offers a whole host of possibilities unachievable by conventional methods, it seems irrelevant to me to add the process of casting (which comes with its own restrictions), especially since it is now possible to sinter directly in precious metals like gold and titanium.

The fineness of the printing determines the quality and appearance of the objects, how strong parts are, print time and cost. The quality from different systems is variable from fine to coarse, dependent on the nozzle size used for extrusion; sheet thickness and grain size. Getting this right is especially important when the object is the final prototype or, like Farah's neckpiece units, Lionel Dean's pendants, and my jewellery, is the final product, no more processes, just finishing.

Layering

An older 3D print method from the 1990s re-configured by Mcor, is that of 'layering', whereby sheets of standard A4 paper, plain and recycled, are inserted into the build chamber one at a time, the profile cut (with a fine, tungsten-carbide blade) and fused/glued to the layer below to build up into the object, hardened and sanded, and the finished product resembling carved wood. Amazingly beautiful objects can be created by feeding the machine with coloured paper. (See Hector Serrano's miniature 'Albert' chair.)

Mcor's Iris 3D printer colours objects in the most obvious way: an enclosed standard 2D colour printer preprints the appropriate profile edges before the ream is fed into the cutting chamber. In a deal struck with Staples Printing Systems Division, they launched a new 3D printing service for 2013, giving wider access to realistic colour in 3D printed 'paper' objects.

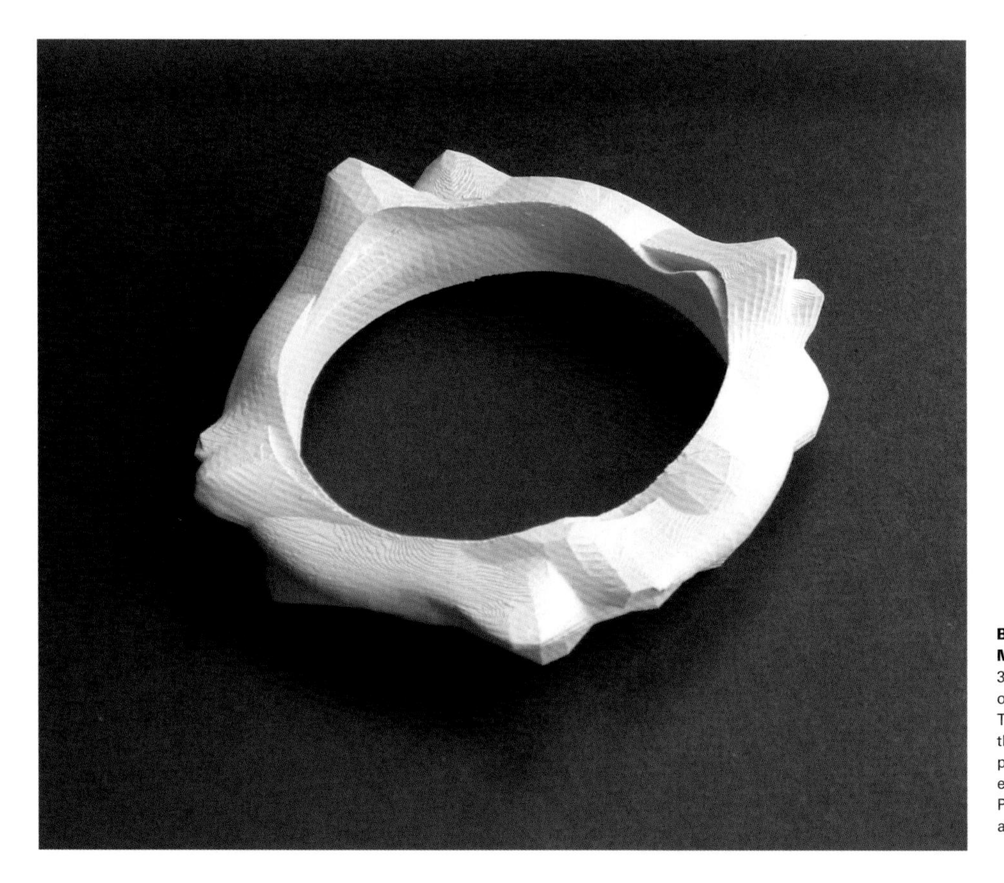

BANGLE BY ANN MARIE SHILLITO. 3D printed in sheets of A4 paper by Mcor Technologies Ltd on their Matrix 300 3D printer and robust enough to wear. 2010. Photographer: the author.

'GROW YOUR OWN BUBBLES' BLUE NECKPIECE BY FARAH BANDOOKWALA. 3D-printed polyamide (nylon) and in stainless steel, Dye and rubber cord. 2010. Photographer: Mido Lee.

SELECTIVE LASER MELTED BRONZE RING BY ANN MARIE SHILLITO, 3D-printed by ES Technology using CONCEPT Lasers machine. Supports removed with minimal clean-up necessary. Partially finished showing the fine quality of this type of printing. 2012. Photographer: the author.

Selective laser sintering

Selective laser sintering (SLS) has been the 3D-printing system favoured by companies, providing a more 'sausage-machine'-type service for consumers. With SLS, models can be complex as the form is supported by the surrounding packed un-sintered material as it is layered. Work can also be stacked to maximise use of the 3D print chamber. Most of the 3D-printed work illustrated in this book is made this way and in polyamide. As the range of materials that can be sintered expands, so too do the applications.

In April 2012, Sculpteo 3D-printed edible macaroons using their special mix of flour, sugar, eggs and almond powder. A bit extreme, but more appealing than the edible insect protein paste mix 3D-printed into aesthetic designs for 'Insects Au Gratin' a themed exhibition in 2012 about all things edible at the Science Gallery, Dublin. The team at the Science Gallery was led by designer Susana Soares, with bioscientist and 3D-printing expert, Dr Peter Walters from UWE Bristol's Centre for Fine Print Research. EU funding for research projects into using insects as a source of protein has ramped up and 3D printing latticed delicacies is a novel disguise to encourage consumption.

Polyamide

Polyamide (e.g. nylon) is the most popular material with designer makers for producing work, whether bespoke, made to order, in series, or in quantities to fit demand, as the finish is fine and the material durable as an end product.

Alissia Melka-Teichroew uses SLS because it combines the strength properties of polyamide with fine resolution at 0.1 mm, which can produce

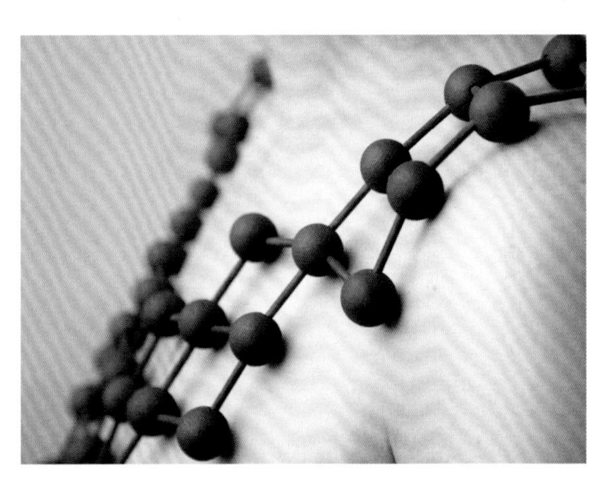

'JOINTED JEWELS' BY ALISSIA MELKA-TEICHROEW. 3D-printed using selective laser sintering, one of the few technologies capable of producing a ball joint in a single operation. Photographer: Lisa Klappe.

NAIM JOSEFI'S FASHION SHOW WITH MODEL WEARING HER 'MELONIA' SHOES **3D-PRINTED BY MATERIALISE IN POLYAMIDE.** 2011. Image courtesy of Materialise.

a functioning ball and socket joint in a single operation. She collaborates with Freedom of Creation to produce her 'Jointed Jewels' collection in different colours, scales and arrangements. She says of her process:

We are able to quickly prototype designs as well as able to create finished and customised objects. The prototyping is very helpful for some of the client work we do.

In 2010, Naim Josefi of Beckmans and Souzan Youssouf of Konstfack, who were at the time students at Stockholm's two most prestigious design schools, collaboratively designed their fully-wearable Melonia Shoes, 3D-printed in polyamide by Materialise for their Stockholm Fashion Show. Jamie Milas at Materialise loved the shoes and wore them at the US RAPID Conference and Exhibition. They were a huge hit and went on to feature on Lisa Roberts' 'My Design Life' with the .MGX collection by Materialise. After a VIP preview screening her CEO, Wilfired Vancraen, heard that her feet did really hurt when wearing them and had a golden pair customised for her as a wedding gift. They had arch supports built in, a special flex structure under the sole to be more practical and were customised with Jamie's name and her husband's on the heels. They also matched her wedding dress. The Melonia shoes were featured at the Victoria & Albert Museum in London in 2011 at 'Industrial Revolution 2.0'.

ZCorp 3D print

The structure of a ZCorp 3D print is a strong but porous matrix, which can be infused with different liquids to provide a range of properties. Resins when cured impart hardness, be machined, used as masters for moulds for both blow moulding forms and thermoforming for fast, efficient prototyping.

SPECIAL 'MELONIA' SHOES 3D-PRINTED FOR JAMIE MILAS FOR HER WEDDING. DESIGNED BY NAIM JOSEFI AND SOUZAN YOUSSOUF, customised for Jamie and Sam by Stijn Derijck and Jari Pallari from Materialise. Selective laser sintered with polyamide and gold paint. 2011. Photographer: Robert Hanet.

This powder-based system can also be hacked. Three of the University of Washington's Open3D students (Meghan Trainor, and later Juliana Meira do Valle and Kate Lien) developed 3D printing in 'wood flour' (black walnut shell, pecan shell, wood bark). The recipe is 4 to 5 parts wood or nut flour and 1 part UF glue for anyone interested!

Sintering fine powdered metals is a major advancement. Steel and bronze have enabled Bathsheba Grossman to give up her day job and sell her mini-sculptures and jewellery via the Shapeways online shop. The structure of sintered steel is porous, so bronze is infused for a solid non-porous structure although it can cause skin irritations and redness if not coated to protect delicate skin.

Farah Bandookwala says:

Different RP materials are combined in individual pieces. I really enjoyed the process of creating organic forms, which appear to have grown rather than been manufactured. Exploring the organic made me think about grafting, and ultimately resulted in a series of nylon bangles that have stainless steel and even transparent acrylic sections grafted into the form... This process of combining materials was a learning curve, as it meant that alongside dealing with constructing the form in Rhino, I also had to consider the most durable joining mechanisms, as well as taking into consideration the different physical properties (shrinkage, flexibility, strength) of the materials in combination with one another.

BANGLE BY FARAH BANDOOKWALA, designed on Cloud9 and Rhino. 3D-printed: SLS in polyamide and steel, SLA in resin. Dye. 2011. Photographer: the artist.

KARI'S TITANIUM WEDDING RING BY ANN MARIE SHILLITO 3D-printed (not finished/polished) with prototype 3D-printed in red and mock-up of engagement ring in white polyamide. 2011. Photographer: the author.

RING BY ANN MARIE SHILLITO. 3D-printed (SLM) in bronze by ESTechnologies. 2012. Photographer: the author.

Titanium

3D-printed titanium has been used in aerospace industries since 2000 and later for dental implants and body prosthetics, as fully melted titanium, using direct metal laser sintering (DMLS) or electron beam melting, has excellent material properties. My daughter's wedding ring is 3D-printed titanium, first prototyped in polyamide to make sure everything fitted. For small items such as jewellery, the cost is easily justified when compared to more traditional methods of manufacture, especially as the fine grain structure means that cleaning, finishing and polishing are not arduous tasks.

Selective laser melting

Selective laser melting (SLM) has expanded the options for metal to gold and silver, which are fully melted to give a solid non-porous end product.

Concept Laser's Mlab machine prints both gold and silver, but silver is the most exciting development, being precious enough to carry the costs of printing, designing and finishing when compared with more traditional methods of manufacture such as lost-wax casting, but not so expensive that it is exclusive. Moreover, combining 3D digital modelling and 3D printing positively differentiates and standard practices such as lost-wax casting might not compete on design possibilities and customisation.

As a Royal College of Art MA student, Markus Kayser designed and built a 3D-printing system that uses sunlight as energy and sand as material. Markus lugged his 3D printer version 2 into the Sahara to focus the intense rays of the Egyptian sun through a 'Fresnel lens' to melt/sinter the sand, which as silica, solidifies as glass to create a granular glass bowl. Beautifully poetic, technology aligned 'where energy and material occur in abundance'.

MARKUS KAYSER with his solar sinter kit in the Sahara near Siwa, Egypt, and the solar-sintered sand bowl made by flattening new layers of sand to melt them. The bowl is 3D-printed from a 3D file. 2012. Photographer: Amos Field Reid.

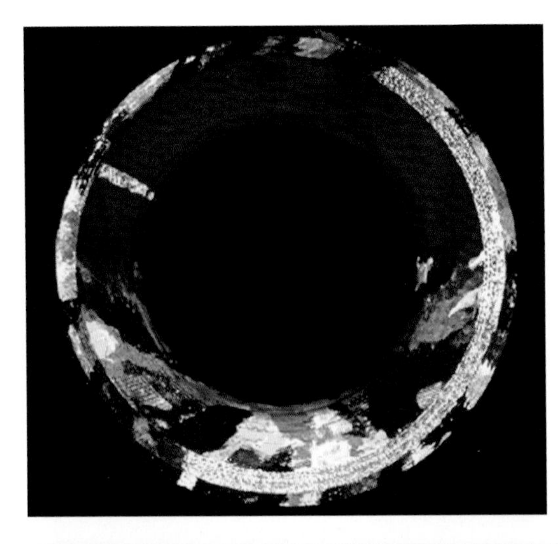

BANGLE BY ANN MARIE SHILLITO, CAD (TriSpectives). 3D-printed in ABS, with three rotatable rings, finished with acrylic paint and gold leaf, 115 mm diameter, 25 mm wide, 1999. Photographer: the author.

OBJECT: AHX 4 (3 VIEWS) BY GILBERT RIEDELBAUCH. CAD, based on a mathematical formula describing a surface. Material: ABS and leaf gold, 3D-printed using FDM system. 227.3 x 203.5 x 181.5 mm. 2001. Photographer: the artist.

Fused deposition modelling

Fused deposition modelling (FDM) systems extrude material through a nozzle to create layers, the standard material being ABS plastic, which can be melted into a very thin strand (as small as 0.04 mm thick) for high quality and surface finish. ABS is robust enough to explore a major advantage of 3D printing – embedded movable parts, so easy to design in CAD, so difficult to construct in real materials but possible with 3D printing.

On a bangle 3D-printed in 1997, when fineness was not an option, the lustre of gold leaf and application of acrylic paint emphasise the textures of the making process and the hatching filling interior spaces. For designer maker Gilbert Riedelbauch:

Stratasys' FDM proved to be the most useful process for me. My area at the university had its own FDM system. As the main user I was able to directly influence the build process. I used the controls of the custom slicing software sometimes layer by layer to achieve otherwise impossible builds.

FDM systems lend themselves particularly well to experimentation. Open-source development of Personal3DPrinters in kit form has flourished since the original patent expired, encouraging all manner of tinkering, converting and subverting by collaborative groups and labs. At the Dutch Lowlands 2012 pop festival, Reiger+Studio and Better Future Factory (in cooperation with TU Delft) presented the Perpetual Plastic Project. Visitors were invited to recycle their plastic disposable cups into rings and other objects by adding their washed and dried cups to a hopper, shredding, melting and finally extruding a material for the Ultimaker Personal3DPrinter to print.

Groups, partnerships and individuals such as Open3dp, Upfront, Jonathan Keep and Hot Pop Factory, exploit the technology for their own purposes. Apart from standard ABS and PLA (polylactic acid, recyclable) an extraordinary range of goo can be extruded – clay, different starches (potatoes), wood fibre, pasta dough, biscuit mix, chocolate, artificial meat and much more not out in public yet! Anything goes.

Esteban Schunemann and Sarah Silve at the School of Engineering and Design, Brunel University, are developing a process termed PDM (paste deposition modelling), and perfecting the extrusion of precious metal clays (silver and bronze) into artefacts that are sintered to pure metal in a kiln. A tessellated pendant designed by Esteban and made using this method was joint winner (with Lynne MacLachlan's 'Phase' jewellery set) of the 'Technological Innovation Award 3D', The Goldsmiths' Company Assay Special Award, at the UK's prestigious annual Goldsmiths' Craft and Design Council 'Craftsmanship & Design Awards 2013'. The award celebrates the innovative use and contribution that technology is making in the craft and industry through the production of 3D jewellery.

EXPERIMENTAL 'PASTE DEPOSITION MODELLING' SYSTEM BY ESTEBAN SCHUNEMANN AND SARAH SILVE. Seashell and ring extruded in BronzClay and PMC Pro. Objects and tessellated ring designed using 3D Studio Max and CorelDraw, processed for 3D printing using Slic3r and CamBam. Photographer: E. Schunemann.

AWARD-WINNING 'TESSELLATED PENDANT' BY ESTEBAN SCHUNEMANN. Designed using 3D StudioMax and CorelDraw and processed using CamBam software. Metal clay (PMC Pro) and its 3D deposition using PDM system. 2013. Photographer: the artist.

3D PRINTED KLIF EARRINGS BY HOT POP FACTORY. Designed using Rhino/Grasshopper. 3D printed in ABS Plastic. 2012. Photographer: May Wu.

Personal3DPrinters (P3DP)

'Personal3DPrinters' (P3DP) are low-cost pre-built and 'build-it-yourself' 3D printers. Most use extrusion to build in ABS or PLA and are spin-off developments based on the original RepRap 3DPrinter project's open-source instructions. RepRap stands for 'Replicating Rapid-prototyper' and the ethos of the community is to print parts for friends and colleagues, spreading 'fabbing' and making it accessible and developable. Founded and invented in 2005 by Dr Adrian Bowyer, Senior Lecturer in mechanical engineering at the University of Bath, UK, he aims to keep RepRap open source and encourage development towards 100% replication of printer parts.

Assembling all the different bits, the open-source Arduino electronics, setting it up, loading the different software to get it to function and tweaking it to print well does takes a geeky mentality. I cheated with mine and bought a pre-built Maxit3D. Personal3DPrinters such as the UP3D and Cubify are less hassle and plug and play.

FDM kits have open casing so the process is visible and good for demonstrating the principles of 3D printing to the uninitiated. These low-end systems can challenge designing. For example, the angle of build is limited; if it is exceeded, supports are needed! Systems with duel extruder heads have either two colours or a support material such as dissolvable plastic. Build time is long and if a model is too intricately detailed an FDM model can result in a blobby mess and tower-like structures end up with lots of stringy bits between them.

Elizabeth Armour used the UP3DPrinter, which has software to automatically create a support matrix. Objects from these printers do have an aesthetic and texture all of their own, which fits her work as it is inspired by microscopic spore and fungi forms.

The next level up from Personal3DPrinters are the desktop types, which are valuable for 'quick and dirty' in-house prototyping, and early-stage testing before committing to an expensive final model. The lower cost of laser hardware with sufficient power to fuse granules and solidify resin has encouraged development of laser-based systems such as FormLab's Kickstarter – funded photo-sensitive-resin-based Form 1 printer, Ron Light's 'Sedgwick Open Source 3D DLP Printer', the 'MiiCraft' and so on.

Pricing

Pricing 3D printed models is by volume of material used, not by height alone and certainly not by complexity – more complexity might mean less volume, and hollow and holey is less expensive than a plain, solid, straightforward cube or sphere. Hollow models need outlets to drain off uncured resin or powder, and should have the recommended wall thickness for material being used.

Scaling a model gives a dramatic shift in price. For example, if a 12 cm model's volume is 100 cm^3 the approximate cost is 100 pounds/euro/dollars. By reducing the height to 9 cm the volume is 50 cm^3 and costs about 50 pounds/euro/dollars, i.e. three quarters of the size will cost just half the original. If you go further, 6 cm gives a volume of 12.5 cm^3 costing 12.50 pounds/euro/dollars: one eighth of the original!

As prices vary between companies for the same or similar materials, it is good to shop around. A web service, 3DPrinting Price Check, gives approximate, but real-time 3D printing prices for the main online providers such as i.materialise, Kraftwurx, Ponoko, Shapeways and Sculpteo, covering materials such as ABS plastic, ceramics, flexibles, glass, metals, plaster, plastics, precious metals and sandstone. You just upload your model as an .stl file, or specify its volume, its surface area or its bounding box size – a very useful and helpful service!

A tip from Ponoko is to prototype in the cheapest material first to test the main parameters, i.e. is it makeable, which can avoid expensive mistakes.

This is all very well, but if your model is too complex to prototype in the least expensive system and cannot be scaled down to reduce costs, comprehensive support on a 'one-to-one model' basis with a bureau might turn out less expensive in the long run than the 'hit and miss' of using online service companies.

'SPRINGIN' SPORE BROOCH BY ELIZABETH ARMOUR. Designed using Cloud9, made in ABS plastic on UP3DPrinter. 3 x 3 x 3 cm. 2012. Photographer: Malcolm Finnie.

BANGLE 2 BY FARAH BANDOOKWALA. 3D-printed (SLS) in polyamide with sections dyed in deep colours, hand-polished steel. 2010. Photographer: the artist.

Finishing

The options for finishing 3D prints are varied and depend on the system and material used.

The coloured polyamides provided by online service providers are dip-dyed. Ordering white and dyeing it yourself gives more colour choices and control over quality. Forums and YouTube are great for finding which dyes and varnish are compatible and the best to use for 3D-printed materials.

The bonded powder of models printed with multi-colours will have been impregnated with extra fixative and no further finishing should be necessary unless a glossy finish is required. i.materialise offer 'flocking', a process similar to that which Elizabeth Armour applied to her ring as a finishing option (see right). The process involves applying a special adhesive to the surface, flocking it and applying an electrostatic charge which ensures that the fibres all stand up at right angles to the surface to give a velvety finish and feel. Finishing processes such as polishing and tumbling are part of the service.

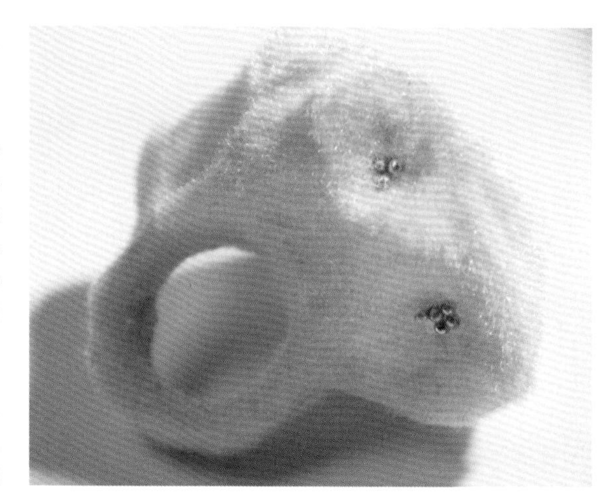

RING BY ELIZABETH ARMOUR. Designed with Cloud9, 3D-printed in ABS plastic on UP3DPrinter and finished by 'flocking', and silver embellishments. 2012. Photographer: Malcolm Finnie.

David Poston had two 3D-printed pieces electroformed with silver. In order to do this, he first had to clean up the sticky cornstarch models, aware that the subtlety of plane changes was essential to the design. He says:

Trying a new technique, or a combination of them, always carries initial risks. I sent the two pieces to be silver plated, expressing concern that there needed to be a sensitive trade-off between plating being sufficiently thick to be durable while taking into account the inevitable loss of detail.

There had been no sensitive trade-off – an expensive waste as at about 190 grams each they were now heavy to wear: small surface nodules would be time-consuming to repair and the refined detail would never be fully recovered. David has seen electro-formed 3D printed pieces done well, in-house by a plating company able to experiment and perfect the product:

Where preciousness is not a requirement, I would now use the print material as the final object, selecting the particular material for its tactile and durability qualities and, if required, the colour opportunities it offered. If I wanted multiple precious metal outcomes I would prefer to electro-form into a silicon mould taken from the printed original rather than plate onto a series of printed originals. The economy of this approach would be significantly influenced by the number of finished products required.

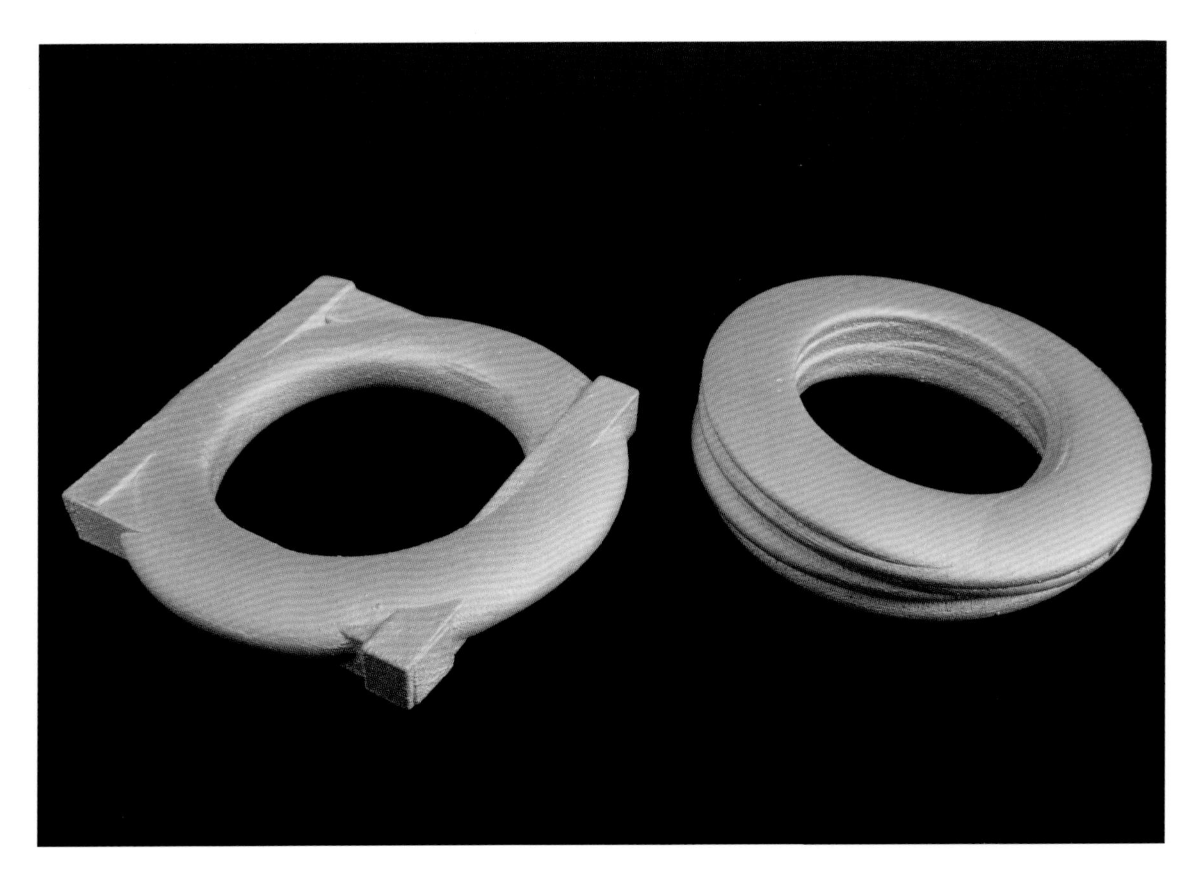

BANGLES BY DAVID POSTON. DrawnReality Project. 3D printed in starch, sealed and electroformed, 2007. Photographer: the artist.

Casting

Jae-won Yoon combines traditional Japanese metalworking technique of Mokumegane with contemporary digital design, his objective being to juxtapose cold, stiff metal and the warmth of the lily pattern. He casts his light 3D-printed structural forms in metal to suspend the precious patterned metal sheets, inspired by natural patterns found within lily leaves, which Mokumegane can recreate accurately. He says:

By using CAD/CAM, I can design detailed and delicate work which is not possible by hand craft alone. In contrast, Mokumegane creates the natural patterns which cannot be expressed by computers. Here, I have combined the advantages of CAD, 3D printing and Mokumegane creating new and innovative jewellery.

His original CAD structure is 3D-printed in wax. Equally, the more expensive photosensitive resins of the SLA 3D print system burn out cleanly and provide a very high degree of detail and finish as master patterns.

Zcorp machines use proprietary 'investment casting material' for 3D-printing parts, a porous mix of cellulose, speciality fibres and other additives, that can be dipped in wax to produce accurate investment casting patterns. Sprues down which the metal flows into all parts of the model need to be added or designed into the 3D print.

In 2011, EnvisionTEC launched a new high-temperature mould material for 3D-printing parts and eliminated the need for casting a metal master. The material is designed to withstand both the heat and pressure of vulcanising the 3D model in rubber, without losing detail or dimensional stability.

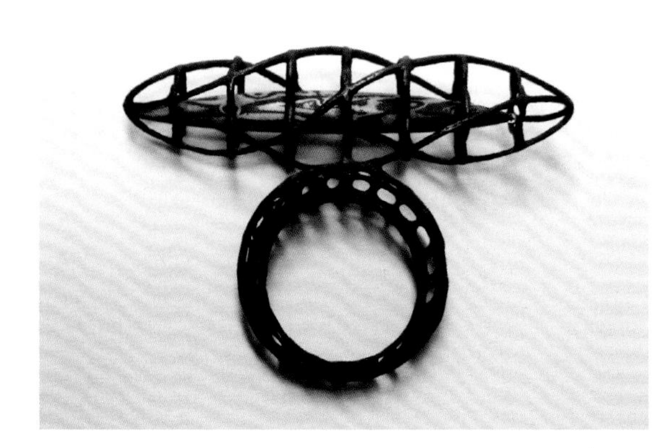

BUD-3 RING AND LILY093 PENDANT BY JAE-WON YOON. 3D printed structures cast in sterling silver, and combined with Mokumegane elements. Photographer: John McGregor.

Another Zcorp product that designer makers can use to good effect is 'direct casting metal material', a blend of foundry sand, plaster, and other additives that produces strong moulds with good surface finishes and is good for directly creating sand-casting moulds as the material can withstand the heat required to cast non-ferrous metals. Somewhere a designer maker must be using this for slumping glass.

Vacuum casting and silicon tooling are economical methods for producing multiple copies of a design where low volumes (around 20 copies) are required.

For his jewellery, Anthony Tammaro uses Zcorp's 3D-printed epoxy-infused gypsum material, EOS's selective laser-sintered nylon, and for the connection components in his larger neck pieces he uses cast silicone units pulled from CNC-machined two-part plastic mould. For higher volumes of parts, as many as 5,000 shots, he could use a urethane mould made by casting around the CNC-machined or 3D-printed master pattern.

Anthony and Alissia, with their 3D-printed neck pieces and bangles laser sintered in polyamide, marketed online, are being joined by growing numbers of other designer makers who are moving towards using digital technologies for their main revenue stream.

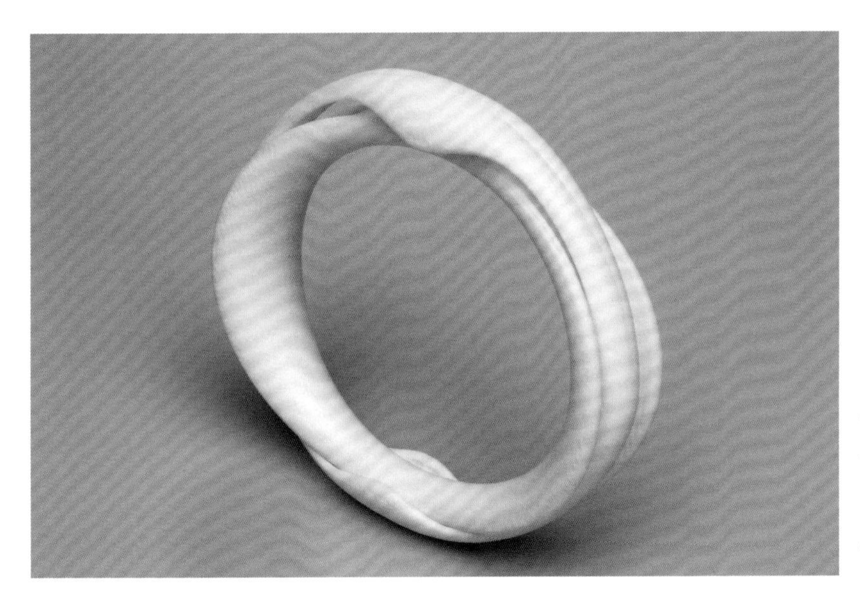

WHITE BANGLE BY ANTHONY TAMMARO.
3D-printed in nylon using SLS system. 2011. Photographer: the artist.

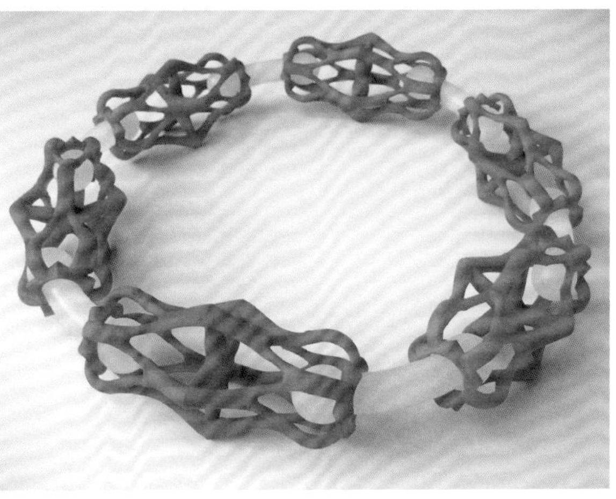

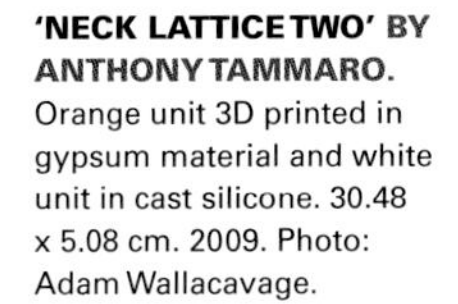

'NECK LATTICE TWO' BY ANTHONY TAMMARO.
Orange unit 3D printed in gypsum material and white unit in cast silicone. 30.48 x 5.08 cm. 2009. Photo: Adam Wallacavage.

'NECK OBJECT ONE' **BY ANTHONY TAMMARO.** Units cast in silicone and 3D-printed in gypsum epoxy material. 342.9 x 76.2 mm 2009. Photographer: Adam Wallacavage.

6 Accessing digital technologies

INTERACTIVE CREATURE SERIES 'QUIVER' BY FARAH BANDOOKWALA for the Jerwood Makers 2011 Award Exhibition. Media: 3D printed nylon (SLS), 3D printed acrylic photopolymer (SLA), electronics, paint. Dimensions: 70 x 26 x 27 cm. 2011. Photographer: David Liddell

There are so many ways to access digital technologies and ways to learn to use them all that it's a matter of understanding and searching for what you need.

Open access to 2D and 3D digital technologies

Farah Bandookwala is a great example of a designer maker who is exploiting different digital technologies to design and produce a series of new interactive pieces. She used Rhino (initially under direction at college and then self-taught), accessed Anarkik3D's haptic Cloud9 software at their office, and had the designs 3D-printed in white polyamide, glass and resin using an online service (Shapeways). She could have used her local Hacklab (like a FabLab but more involved in the electronic side of making things) to learn how to connect up the Arduino electronic board and components such as motors, sensors, LEDs, sound and rumblers, (purchased online) or how to program using 'Processing', but mostly she worked out how to do it herself, using online tutorials and support from friends.

Weston Beamor
Redeye-on-demand
100kGarages
Glasgow Sculpture Studios
EASY, AFFORDABLE, OPEN ACCESS
ACADEMIC CENTRES
LOCAL hands-on workshops FABlabs SUMMER SCHOOLS 3D PRINTING Inge Panneels 3DPrintUK
Jenny Smith
Colleges and Universities
ACCESSING Farah Bandookwala
DIGITAL electronics ADVANCED COURSES
TECHNOLOGIES online access
Gilbert Riedelbauch
UNIVERSITY RESEARCH
BEATE GEGENWART
carbon footprint
CHOICE Zachary Eastwood Bloom
PONOKO
laser cutting
milling
Bureaux
Beate Gegenwart
focused courses SCULPTEO
OTTO GUNTHER SERVICE PROVIDERS
COLLABORATION
MARIKO SUMIOKA
MetropolitanWorks
Shapeways
Katie Bunnell
ShopBot
Staples Easy 3D
MICHAEL EDEN
COLLABORATIVE RESEARCH
MATERIALISE
HAND HOLDING SERVICES
Gordon Burnett
LaserLines

Through local FabLabs you not only gain access to an array of flexible, computer-controlled tools and other DIY equipment, you are accessing an inventory of capabilities, a global network of technical support and collaborations which enables almost anything to be made and invented. The labs, continually growing in number, are all linked internationally – all this initiated out of the MIT lab and into a major movement. Scotland had its first, the MAKLAB, in 2012 (see General links and websites on pages 157–8). At these labs you learn to do it yourself and to share: designer makers should join with their valuable knowledge and skills. The reward could be the realisation of an idea into a new range of work.

Other membership-based establishments where there is access to equipment and resources include Glasgow Sculpture Studios and Metropolitan Works in London. The Works also aims to educate and promote the take-up of digital manufacturing, holding seminars and advanced courses and providing induction and training into workshop practice for more hands-on use of equipment and machinery.

Short, specialist courses

Short, intensely-focused courses, workshops and summer schools for small groups are ideal and effective for learning more about software programs and methods of production such as laser-cutting and 3D-printing that will fit your practice. It is inspiring to work with like-minded people for a day or two. To give a few examples: as mentioned briefly in Chapter 4, Edinburgh Laser Cutting Studio artist Jenny Smith holds training courses where you can learn the digital skills necessary to prepare and laser-cut your own work. Courses cover generating files digitally, converting hand-rendered imagery into digital files, cutting or etching from photographs, and with supervised access to machinery you acquire the technical skills to use individual settings and materials, and to get the best results. Shapeways website has a forum that promotes short courses to learn 3D modelling (usually Rhino) for 3D printing.

DO I NEED DIGITAL TECHNOLOGIES?

Before investing in digital technologies you should review the following checklist:

- What type of work do I do now? What will I be doing and what will I need to do in the future?
- Will using digital technology enhance my practice?
- Do I need my designs in a digital format?

For Tavs Jorgensen, digital technologies enable new creative opportunities and approaches and can inspire new sets of ideas. Software he uses includes Rhino, Illustrator with CADTool, AutoCAD and Magics.

For Dutch fibre/textile artist, Monika Auch (who uses a specialised weaving program), and for researcher Cathy Treadaway (Freeform, ZBrush), the benefits of their software are that they speed up processes and provide them with new 3D possibilities and new 3D forms.

- Do I need computer-aided designing?
- Do I need 2D or 3D technologies, for laser-cutting or 3D-printing, or both?

Johanna Spath and Johannes Tsopanides adopted digital applications (Rhino, Cinema 4D, Processing) into their practice and new business for the excellent prospect of materialising unique, individualised digital 3D models.

- Do I need to exchange design information electronically with manufacturers and suppliers?
- Do I need specific formats for production such as decals, vectors, geometry?
- Do I need to exchange data between different software programs?
- Which applications support the range of formats I will need?
- Will my designs transfer smoothly into other formats and applications for different output, such as standard 'colour' conventions for printing, .stl for 3D-printing?
- What level of complexity do I actually require. What can I cope with?
- Do I need good realistic-looking presentations to liaise with customers and clients?
- Do I need to also spend time learning to use render functions or another package, with stuff such as ray-tracing for reflections?
- Are the basic graphics in programs such as SketchUp and Cloud9 acceptable?
- Which applications can be ramped up if and when I need more advanced functionality?
- Should I choose a free version or a Pro version?
- Is my current computer or laptop suitable and even compatible with what I am aiming for?
- Mac or PC? (Industry is mainly PC-based, so is most CAD software. Some like Rhino are cross-platform. Macs dominate graphics. Some CAD software is for Macs (Form Z and Deskartes). Some open-source applications are linux-based.)
- Will I need training? Does the software supplier/vendor provide this and a good backup system, if I need these?
- Should I collaborate or pay someone experienced in CAD to work with me or for me when I need it and will I get what I want?
- Is there a workshop, FabLab community or Techshop nearby, or a college with open facilities?
- Are there other ways to get this data, e.g. draw, sketch and scan or get it digitised?
- Where do I find information, what search terms will be helpful to use on the Internet?

Types/examples of software for designing:

2D designing
Illustrator, CorelDRAW, Inkscape, 2D Design

2D visualisation
Photoshop, GIMP

3D visualisation
3DStudio, SketchUp, Cinema4D, Maya

3D CAD (engineering based)
TinkerCAD, 3d Tin, Blender, Rhino, AutoCAD, Zbrush, Inventor, Solidworks

3D Modelling
Cloud9, Freeform, Sculptris, Maya

Generative software
Processing, Grasshopper, Jenn3D

3D File repair/testing software
Magics, NettFab, MeshLab

Software used by the designer makers replying to a questionnaire and also gleaned from articles for this book, with expertise ranging from medium to higher level:

GR: *FormZ, Bonsai*
KH: *Rhino 3D*
CT: *FreeForm, Zbrush*
DW: *Maya, Rhino*
ZAC: *Rhino, Terragen, Maxwell*
MA: *Norwegian program: special weaving software*
LD: *AutoCAD*
JSP: *Rhino, Cinema 4D, Processing*
TJ: *Rhino (includes tools for scanning and point cloud data, Illustrator with CADTool, and also a little AutoCAD, Magics*
A: *Illustrator or SolidWorks*
AB: *Photoshop*
BGr: *Rhino, Mathematica, Surface Evolver TSplines for Rhino, ZBrush, Grasshopper, Magics, MeshLab*
AMS: *Rhino, Cloud9, Teamwork*
FB: *Rhino, Cloud9 and Processing*
RB: *Grasshopper, Rhino*
LM: *Jenn3D, Grasshopper, Rhino*
BLUEM: *Inventor*
AC: *CorelDraw*
BGA: *Photoshop, Illustrator*
GM: *Rhino ('I like Rhino because it lets me make mistakes')*
IP: *Rhino, Illustrator, Photoshop*
JSM: *Illustrator*

I also run one- to three-day courses throughout the year for designer makers to learn how to use Anarkik3D's haptic sketch modelling software, Cloud9. The objective of these courses is to understand, and to be able to design for, 3D printing and apply these technologies productively within designer makers' practice.

▲ **JEWELLERS HILTJE AND BIRGIT** using Anarkik3D's haptic 3D modelling software, Cloud9 with touch enabling 'Falcon' device, on their 3-day course in 2012. Photographer: the author.

Colleges and universities: access and collaborating

Going back to college is another option, as this is where most designer makers have learnt to use digital technologies. Zachary Eastwood-Bloom became computer proficient as a student at college. Lionel Dean's FutureFactories began as a blue-skies university research project in 2002, where it quickly expanded into a PhD thesis and then into regular studio practice. Johanna and Johannes and several others (Lynne MacLachlan, Riccardo Bovo etc.), have or have had access to computers and a range of software, cutting and 2D/3D printing equipment, plus the technical expertise, as students with time to learn. Teaching staff at universities and colleges, like Inge Panneels and Beate Gegenwart, also have access to digital technologies to explore their niche areas.

'FRAGILE COOLIMONS': THREE BOWLS BY GORDON BURNETT. 3D printed in nylon. Largest bowl: 250 x 325 mm. Owner: maker. 1999. Photographer: Stuart Johnston.

Beate notes:

It is useful to work at a university as we have many lasers (for paper, textiles, acrylic etc) and a waterjet cutter. I also received research funding from our university, ESDF and Arts Council of Wales funding; all of course making the development of work much more realistic.

A year at Monash University, Melbourne, Australia, resulted in Gordon Burnett absorbing complex cultural issues and being more creatively fluid within the digital environment, learning 'Maya' software, and having the luxury of unrestricted access to an early 3D-printing machine. Symbolic functional objects were 'grown' digitally then cast in metal, glass and ceramic: touching them, feeling their weight, sensing their reality once out of their virtual world is critical for assessment and this engulfed Gordon in a creeping obsession to control and impose his will on the partnership with technology.

In Sept 2012 Michael Eden was appointed as a Research Fellow at MIRIAD, Manchester Metropolitan University to focus on 'Future Forming and Digital Technology' and with access to funding opportunities and resources will continue developing his extensive specialist knowledge about designing for and 3D-printing ceramics.

■ COMBE DOWN STONE MINES PROJECT: '1479 PLATES'

'1479 PLATES' BY CHRISTOPHER TIPPING. Detail of 10" bone china plates with digitally designed surface decoration produced in collaboration with Katie Bunnell. 2009. Photographer: the artist.

Christopher Tipping, an artist with a background in ceramics who works exclusively on site-specific projects collaborated with Dr Katie Bunnell at University College Falmouth to realise '1479 Plates', a commission from Bath and North East Somerset Council to commemorate the Combe Down Stone Mines Stabilisation. His concept explores the relationship between present-day engineering and mining technology, stone mines heritage, natural history and two eighteenth-century entrepreneurs, Ralph Allen and Josiah Wedgwood.

When I asked Katie to comment about '1479 Plates' for this book she highlighted the creative potential for engaging with digital technologies through collaboration – particularly with a skilled digital craft person, digital designing, hand drawing and the exploitation of digital ceramic print technology for 'one-off' production, which mapped to Katie's own research.

Christopher: *Collaboration within my practice is very important to me. My hands may have become almost obsolete as a making tool. My making was being done in factories ... un-involved with the process. What I got from Katie and Autonomatic, was collaborative dialogue and creative empathy in a surprisingly un-technological and personal manner... My own perceived lack of digital skills put me firmly on the sidelines. Client teams often work exclusively in a digital world – but not in any manner ... in which I thought or produced ideas – by engagement with materials, touch and bespoke craft process. I was beginning to feel distanced from making. I realised quickly whilst working with Katie that I was simply not speaking the right language and had actually developed ... a fear factor about digital craft processes. Katie's own use of technology, hardware and language is remarkably and refreshingly un-techie. She quickly understood my process and began ... to tailor the digital resolution around this. Perhaps we were fortunate with regards to personality. I do not underestimate that this was the outstanding factor in the success of the work.*

For more information about Katie Bunnell and a link to Autonomatic, see Contributor biographies and contact details for designer makers, pages 149–52.

'1479 PLATES' BY CHRISTOPHER TIPPING. Installation of 788 bone china plates showing the artist's interpretive map of Combe Down Stone Mines. 5 x 9 m. 2009. Photographer: Kevin Fern.

Katie: *Christopher would describe verbally, through visual references and sketches, what he wanted, I would try things out... Through this emergent and discursive process the map began to take shape. There were periods of creative frustration ... as things developed he quickly began to see that even CAD has limitations! There was also a sense that Christopher wanted to be more hands on, to do it himself, but had to rely on me as his interpreter or cipher.*

The use of CAD software enabled us to produce a work on a vast scale and to create a visual vocabulary for the artwork. The capabilities of ceramic digital decal printing enabled Christopher to imagine and realise a wall of 788 individual dinner plates, which were printed and fired onto bone china by Digital Ceramic Systems in Stoke-on-Trent.

On reflection, Christopher suggests that if he had known more about the process, the time involved, dialogue required, the importance of being in the same space at the same time, he would have been more prepared, developed more sketches, more design work, leaving more time for experimentation and sampling. Importantly, he recognises that the intensity of the dialogue and the speed of the development of the map towards the project's deadline were integral to creating a dynamic and exciting piece of work.

Christopher: *There is now incredible added value to my processing of new concepts and ideas for projects as I filter the options available to me through that pivotal experience with Autonomatic.*

Academic centres support collaborative research and development projects with outside practitioners. Katie Bunnell (see opposite panel) leads the 3D Digital Research Cluster at University College Falmouth, which includes designer makers Justin Marshall and Tavs Jorgensen, whose own works are cutting edge.

Jenny Smith accessed laser-cutting technology and expertise at Centre for Fine Print Research at the University of West England, including invaluable one-to-one training funded by the Scottish Education Authority:

I didn't have regular access to a laser cutter, which meant I worked on the machine in short concentrated bursts... Initially I saw this as a hindrance to my practice, but increasingly, I saw it as giving a focus to my working process... I was often forced to make very immediate decisions about whether to go with something or re-do it when a result wasn't exactly as I anticipated. I may not be sure if a piece is working when it initially cuts and often don't have time to evaluate the work as I cut it, because the cutting time was so precious. When I get back to the studio and have time to reflect, I increasingly welcomed the spontaneity of the results that were least expected.

'PARASITE' MAGNETIC BROOCHES BY FARAH BANDOOKWALA.
Designed using Cloud9 and Rhino, 3D printed in 'nylon like' polyamide, and steel by Shapeways, finished by dyeing, polishing and in-setting magnets. 34 x 34 x 21 mm. 2010. Photographer: the artist.

Service providers

Internet-based consumer-facing service companies (Ponoko, Shapeways, i.materialise, Sculpteo, 3DPrintUK, Redeye-on-demand, Staples Easy 3D and so on) provide easy and affordable online access to laser-cutting, 3D-printing and milling to anyone. All you have to do is set up an online account and upload designs for manufacture. Their websites provide not only immediate feedback about makeability, problems they can fix, ones you have to fix, selections of materials available and suitable for your design, the cost, resizing models to reduce cost, etc., but also tutorials, courses and some design software. Once you are happy with your design, you pay them to produce it and some days later you receive or collect your parts.

Farah Bandookwala has her 3D-printed units sent to her workshop to finish with colour, findings and magnets before sending the finished pieces to retail outlets, galleries and exhibitions. Although the 3D-print service is swift enough to work to order, the cost price for printing is low enough to hold small amounts of stock.

It's worth setting up two or three accounts with different providers for greater choice of materials and facilities. What's more, their shops attract different customers. Bathsheba directs hers to her online service provider's 'shop' and it is a very effective money earner – her small sculptures and jewellery are finished by the service provider and sent direct to her customers.

GILBERT, OTTO AND PONOKO

GILBERT'S 'OBRUT LIGHT': base and shade, showing laser cut units from Ponoko, and variations of the finished light. Acrylic, aluminium and polypropylene. Dimensions: 181 x 181 x 40 cm. 2009. Photographer: G. Riedelbauch.

Gilbert Riedelbauch and Ponoko

Gilbert Riedelbach at The Australian National University, Canberra, Australia, fully integrates online-based fabrication by using Ponoko:

...collaborating with external fabricators provides me as a designer/maker with access to a wide variety of materials and processes helping to extend my professional practice. New design ideas and objects are possible through accessing these processes. This collaboration is based on digital vector drawings used as the communication with the online fabricator.

He designed his 'Obrut light' so that shade, base and metal fixture are all laser-cut by Ponoko and the shade snaps into the base forming an illuminated dome to take advantage of LED technology.

He designs in Illustrator, uploads vector drawings to his account on Ponoko's website and is then able to have designs repeatedly produced on demand, reliably and with precision. In his account, he gets feedback on viability, cost, and once the process starts, feedback on progress.

It is essential to successfully integrate these technologies into studio practice. Knowledge of materials and processes (for both traditional and digital making) and his designs, right from early concept stage, are informed by understanding the advantages and limitations of collaborating online to access remote or distributed processes for fabricating.

For Gilbert, this marks a new level of collaboration. The power of this system is his taking control of the fabrication equipment and the machine's tool-path via what he visualises on his computer screen: vector drawings (machine code where mathematics is the universal language at its centre). On reading his file, a laser cutter springs into action and translates the design into real parts.

Otto Gunther and Ponoko

As a digital fabricator Ponoko has been instrumental in enabling Otto Gunther to develop his work and style through the freedom to create on a much grander scale, to dream big and still maintain financial sanity:

While nothing beats owning your own laser cutter, CNC machine, or 3D printer, the expense is still outside most artists' budgets.

In striving for mathematical authenticity, Otto predominantly uses laser cutters to engrave his designs onto different mediums; the accuracy of this method appeals to his compulsion for perfection. It is the intersections in art that Otto finds most interesting, so he looks to create in ways that will find these. Although he starts with a direction in mind, his design approach is about exploring and manipulating shapes by combining, contrasting and bringing together by hand the cold efficiency of machines with the warmth and randomness of nature. His designs start in Adobe Illustrator and this allows him freedom over traditional mediums to explore and create movement through 2D geometric shapes, massaging designs until they start exhibiting the type of motion Otto is looking for. Once the design is established he walks away for an extended period. If it still speaks to him when revisited with fresh eyes, he works on colour choices, type of medium and the overall size of the piece.

Services like Ponoko's help to streamline the manufacturing aspects of Otto's process, although 80% or more of his time is spent hand-finishing each piece, hand-inking and completing design elements (such as building the frame), and finally finishing the wood to enhance its beauty and the design.

OTTO'S 'RADIAL' PANEL. Designed using Adobe Illustrator CS5. Laser etched onto wood, cherry and birch veneer, hand inking (acrylic ink) and finishing. 39.4 x 80 x 5.7 cm. 2011. Photographer: O. Gunther.

Other examples of service providers include:

Ponoko, a New Zealand-based company, set up at first to implement an online fabrication system, which is now an international network of hubs for laser-cutting, CNC-routing and 3D-printing. Gilbert Riedelbauch and Otto Gunther use Ponoko to outsource parts of their production (see opposite panel).

100kGarages, set up in 2008 by ShopBot in the USA in collaboration with Ponoko, is a 'global place' connecting those with the designs and ideas with digital fabricators ('Fabbers'). 100kGarages is not a 'middleman' – it is a facilitator depositing designers' designs at the nearest appropriately resourced 'garage' to keep the carbon footprint as small as possible. The designer has direct contact with fabricators to negotiate manufacturing their work.

Staples Easy 3D, a new online 3D-printing service from Staples Printing Systems Division, who have teamed up with Mcor to offer a simple uploading system for electronic files and to pick up your model.

The Colour Company is the first 3D print company to bring 3D-printing to the high street in London or the UK and making it more accessible to drop by and collect your work.

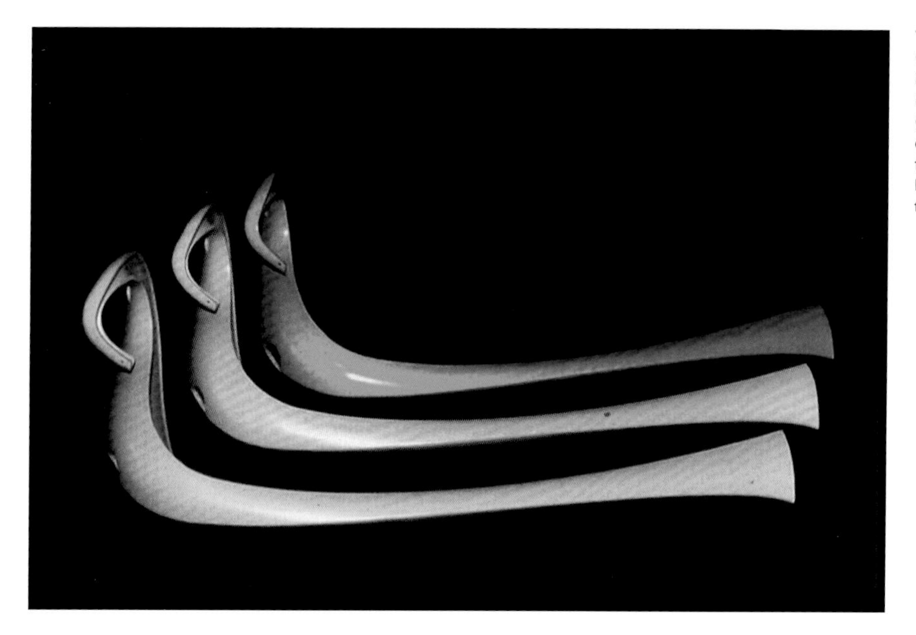

THREE LEG UNITS FOR COFFEE TABLE BY ANN MARIE SHILLITO. Rhino. Prototype 3D-printed (stereolithography) and vacuum-cast from silicon mould made from prototype. Made at Materialise. 2000. Photographer: the author.

Professional services and bureaux

Consumer-facing service companies work within parameters that suit them, i.e. with one or two variables, whereas professional services and bureaux can work and advise on a one-to-one basis to maximise on properties that parts need to have, and recommend the best system. Professional service bureaux specialise in one or more systems and allied technologies such as scanning models, prototyping (advising on the right material – durable enough to withstand the forces of moulding), mould making, casting, various finishing methods and hand-holding support. Examples of such services include Materialise, who, with their commitment to innovation and high-quality support (anyone and any company can create world-class products and Materialise are there to help them); and Weston Beamor who offer services to help every jeweller or retailer (from students to high-end designers) with their particular manufacturing needs.

I used Materialise to get six 'legs' vacuum-cast from a mould – made from a 3D-printed prototype – made from my Rhino design of a coffee table and to make two tables, and to bureau LaserLines for the flexible neckpiece for the 'Interface Exhibition' in 2005.

COFFEE TABLE BY ANN MARIE SHILLITO using Rhino, vacuum-cast from 3D-printed prototype. 2000. Photographer: the author.

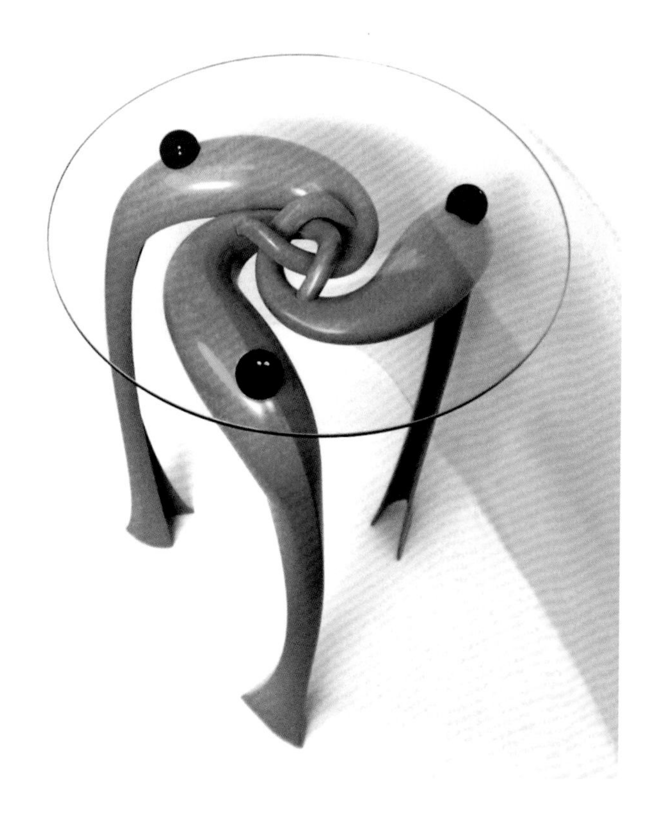

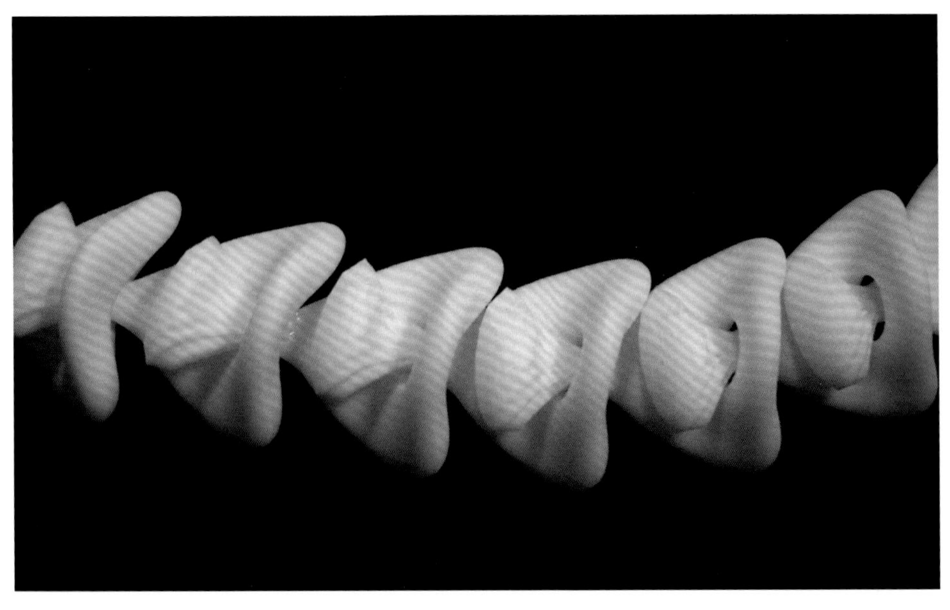

NECK PIECE FOR 'INTERFACE EXHIBITION' BY ANN MARIE SHILLITO. Material: ABS, 3D-printed by Laserlines. 2005. Photographer: the author.

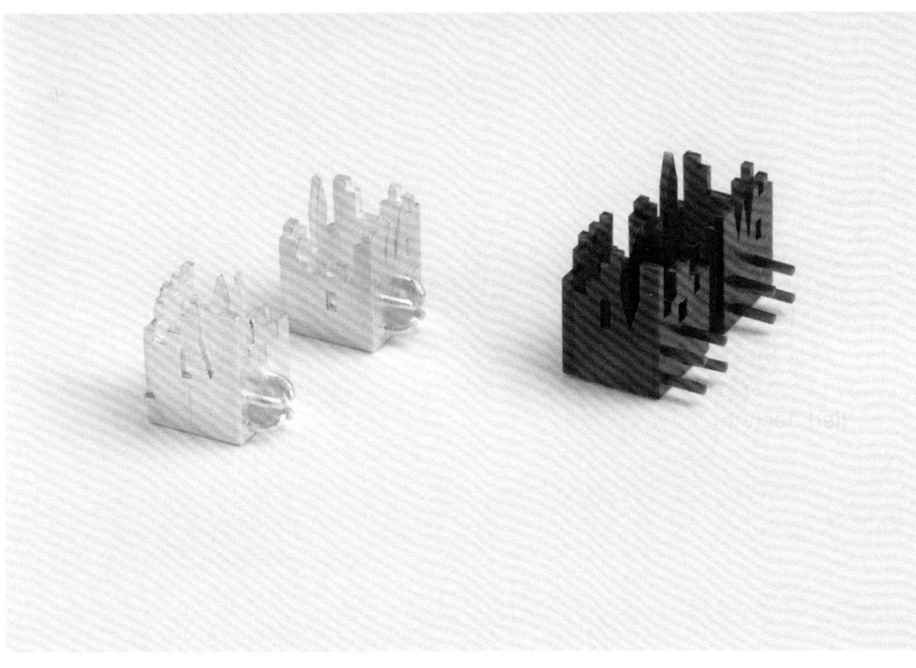

JEWELLERY UNITS BY MARIKO SUMIOKA. ECA College Project: CAD design. 3D-printed by Weston Beamor, and lost-wax cast in precious metals. 2009. Photographer: John K. McGregor.

Jewellery students at Edinburgh College of Art are inducted into digital manufacture using Weston Beamor's services, getting guidance on CAD files for 3D-printing, casting and hallmarking (if required), all under one roof to simplify and speed up the process, guarantee good results and prevent expensive mistakes.

7 And finally...

FREE DESIGNS FROM THIS BOOK'S OWN SHOP SPACE AT I.MATERIALISE TO PRINT IN 3D, from the smallest (three of which have entrapped and 'freely' moving centre pieces):
1. pendant/key fob;
2. lacy pendant;
3. larger pendant/key fob (with two parts entrapped within each other);
4. a spiky globe;
5. latticed/lacy Globe for LED tealights.
Photographer: Siri Rodnes.

In the introduction I quoted J. R. Campbell and Malcolm McCullough who believe that designer makers have the 'right approach, skills and mindset' to explore the 'close relationship between digital work and craft practice'. The objective of this book is to inspire and motivate participation in digital technology by understanding it better from a designer maker's perspective. I hope that this book at least facilitates the first steps.

Want to engage with 3D printing?

For non-CAD users who now want to engage with 3D-printing technology (which is attracting more and more attention), I have set up an account at the service company i.materialise. This is the book's own 'shop window' and it is available at http://i.materialise.com/designer/ann-marie-shillito, or by scanning the QR code opposite with your Smartphone. It has 'free-to-use' designs that can be 3D printed by i.materialise in a choice of materials and prices (a small percentage of which covers admin and acquiring further designs).

specialist software development
award winning designer makers
COLLABORATIVE PROJECT
MORE INTUITIVE
Suzanne Esser
Michael Eden
shape changing
TAILENDS
'FREE TO USE' DESIGNS FOR 3D PRINTING
UNIVERSITY OF HUDDERSFIELD
SOFTWARE PLATFORM
AUTOMAKE
Masako
University College Falmouth
HAPTICS
Birgit Laken
FORCE FEEDBACK
prototyping
Creative Commons Licence
GRAPHICAL USER INTERFACE
QUICK RESPONSE (QR) BARCODES
virtual 3D touch
wireform mode
CO-DESIGNING
Elizabeth Armour
CLOUD9
FARAH
SMART phone barcode reader app.
Freedom of Creation
UCODO
STORY TELLING
Jerwood Makers 2011 Award
Lionel Dean
BOOK'S OWN 'SHOP WINDOW'
HILTJE WYNIA
COMPUTER-BASED GENERATIVE SYSTEMS
Justin Marshall
Kari's wedding ring
shop at i.materialise
Assa Ashuach
EXTRUDED DESIGN
non-CAD users
Bandookwala
3D PRINTED IN TITANIUM
Hamaguchi
NEW TO 3D COMPUTER MODELLING
Novint's Falcon haptic device
3D INTERACTIVE TECHNOLOGY
Anarkik3D
Less intimidating
co-creation experience

Ordering is straightforward: add the address for shipping and pay. These designs are provided under a Creative Commons Licence – Attribution Non-commercial Share Alike, which means you can use the 3D prints as you choose for non-commercial purposes only, you must credit the original designer and you must license any new creations under identical terms. Enjoy your 3D prints.

The adjoining panel highlights co-designing, which is another great way to get involved with digital technologies!

Haptic 3D modelling for non-CAD-using designer makers

I mentioned in the introduction that I am a co-founder of Anarkik3D, a specialist software development company that uses 'haptics' (virtual 3D touch) as a better way to interact digitally in three dimensions. Our 3D-modelling product, Cloud9, is designed specifically for non-CAD-using applied artists and designer makers to more easily create in 3D for 3D printing. We exploit haptics technology for a more intuitive interface, as haptics utilise one of our fundamental senses, touch and feel, providing a more familiar way to interact digitally in three dimensions. To touch something is to understand it – emotionally and cognitively. By tapping in to our tacit knowledge of what we know of real-world interactions and an object's materiality, even the low-level sensation of touch provided by the device is valuable as it

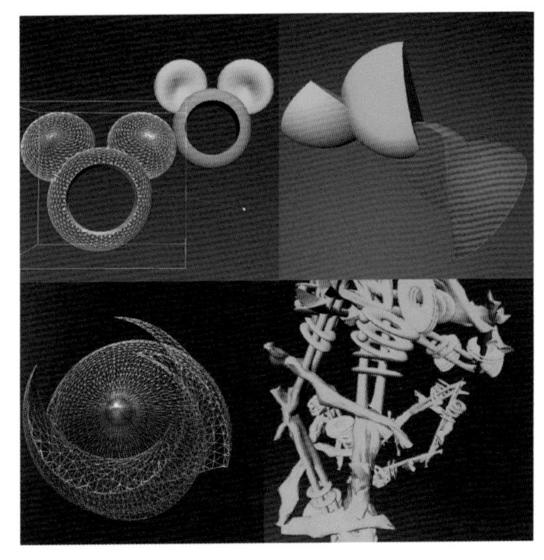

◀ **SCREEN CAPTURES:** models created in Cloud9 3D-modelling software. Clockwise from top left: 1. Birgit's fun ring; 2. Hiltje's first time ever designing and modelling in 3D – slicing, rotating, colouring and moving; 3. Masako creating amazing complexity by simply using functions such as one-axis scaling, deforming, move, rotate and mirror; 4. Suzanne's two Boolean parts in wireform mode, experimenting with subtracting a torus from a cone to create beautiful, interesting, complex forms. 2012. Screen captures: Anarkik3D.

BIRGIT LAKEN'S CLOUD9 RINGS: 3D-printed – two in polyamide (one as it comes and one dyed) and one in alumide. 40 x 37 x 10 mm. 2012. Photographer: the artist.

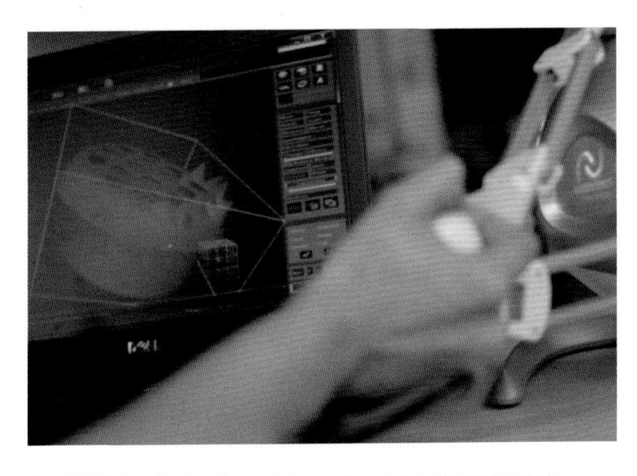

Anarkik3D's bundle: haptic modelling software, Cloud9 and Novint's Falcon haptic device. Photographer: David Liddell

'CORAL' BROOCH BY ELIZABETH ARMOUR. Part of her 3D-printed 'New Breed' Collection, displayed at Kinectica Art Fair 2013. ABS plastic, sterling silver. 6.5 x 5.3 x 5.2 cm. Photographer: the artist.

renders the virtual less intimidating and facilitates design tasks like manipulating, moving, and rotating objects and virtual space.

Cloud9 is modelling software bundled with a haptic device and developed from academic research where I was involved in leading a team of artists and computer scientists who were looking into the potential of haptics as a more coherent interface for working in virtual 3D. The 3D haptic device provides force feedback and movement in three axes, which makes navigating the 3D environment more straightforward and the GUI (Graphical User Interface) less complex than CAD, and easier and more fun to learn and use, enabling users to be creative from the first day they use it.

These models are by four jewellery designers (Suzanne Esser, Birgit Laken and Hiltje Wynia from Holland, and Masako Hamaguchi from London), all novices to 3D-computer modelling, who all created 3D-printable signature models after taking a three-day intensive course on Cloud9.

Two other Cloud9 users included in this book are both award-winning designer makers: Farah Bandookwala, who won the Jerwood Makers 2011 Award to design, 3D-print and construct four interactive sculptural pieces, and Elizabeth Armour, who won the F&A Bradhsaw Award as a graduate, which she used to purchase Cloud9, also to design, 3D print and construct jewellery.

CO-DESIGING

'SUPERKITSCH' JEWELLERY BY LIONEL DEAN. Charms selected from 'Superkitsch Library' and 3D-printed in polyamide. 2011. Photographer: Anwar Suliman.

SHAPE CHANGING LEMON SQUEEZER DESIGNED BY ASSA ASHUACH for co-designing and for 3D-printing. 2011. Courtesy of Assa Ashuach Studio © Assa Ashuach Studio.

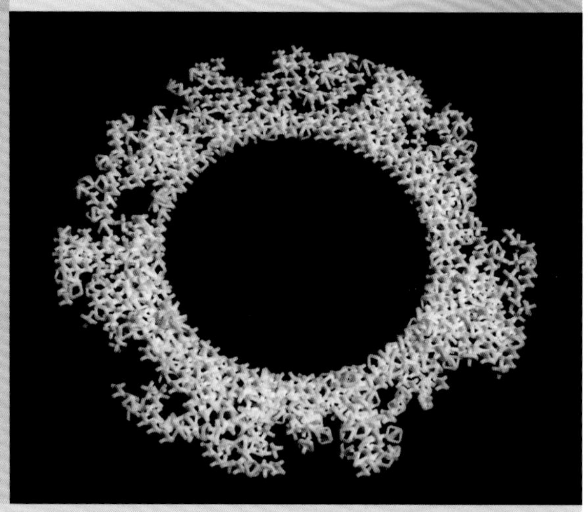

'RANDOM BRACELET 2' BY JUSTIN MARSHALL. Selective laser sintered (SLS) nylon, rapid prototyped, 60 x 160 x 5 mm, in the collection of the National Spanish Design Museum (DHUB). 2009. Photographer: the artist

Co-designing

Lionel Dean's SuperKitsch Jewellery is about a co-creation experience for you to 'create your own' by selecting a set of commonplace, banal, 3D charms from a virtual library, which are then assembled at random to be 3D printed as 'wearables'. A SuperKitsch bolo necklace and ring are available at Freedom of Creation (freedomofcreation.com) and there is the potential for users to submit their own charms for others to use.
Assa Ashuach's 'shape changing' collection for UCODO.com's 'software products' includes his lemon squeezer. 3D interactive technology enables users to customise, change shape, order it online and have it locally produced.

The software platform Automake combines computer-based generative systems so that, with a degree of control from the user, random bracelets are assembled from a range of modular units in an infinite variety of one-off structures. Automake started in 2006 as a collaborative project between Justin Marshall (at the University College Falmouth) and the University of Huddersfield.

STORYTELLING USING DIGITAL TECHNOLOGIES

Michael Eden uses Quick Response (QR) barcodes to explore their potential for storytelling. From the URL of a website page where stories can be told, memories stored, photos deposited, sounds locked away, a QR code is generated for the commissioning client. Using CAD, Michael extrudes this two-dimensional image into a three-dimensional form, which is 3D printed as a unique object.

The unique QR code pattern scanned with a smartphone barcode reader app connects the object, the owner and future generations to the web page where the story of this heirloom can be experienced virtually.

'MNEMOSYNE' BY MICHAEL EDEN. 3D printed, using selective laser sintering, in high quality nylon material with soft mineral coating. 14.5 (h) x 18.5 (w) x 18.5 cm (d). 2011. Image courtesy of Adrian Sassoon, London.

MNEMOSYNE'S UNIQUE QR CODE. Used by Michael Eden for both extruded design and, via a smartphone app, links to its own story and history webpage. 2011.

And finally...

I use my jewellery skills to continually test Cloud9 for usability and compatibility with CAD packages (such as Rhino) and for 3D printing. In 2011, I used Cloud9 to design my daughter's wedding ring, which was 3D printed in titanium, my metal of choice for the past 30 years – it was researching laser cutting titanium that drew me into digital manufacture in the first place. Kari's ring is therefore pure poetics, completing a very neat, personal circle and fulfilling my aspiration to 3D print in titanium – a very fitting wrap-up to this book.

KARI'S WEDDING RING, BY ANN MARIE SHILLITO. Designed using Cloud9 3D-modelling software, and 3D-printed directly in titanium. Kari's hand-wrought gold and diamond engagement ring slots into this wedding ring, tested by prototyping by 3DPrinting rings in polyamide. 2011. Photographer: Siri Rodnes.

Appendices

Contributor biographies and contact details

Elizabeth Armour

Lizzie graduated from Duncan of Jordanstone College of Art, Dundee, in 2012 where she studied Jewellery and Metalwork Design. The focus of her work is designing through making, inspired by the natural world and new technologies to create a 'new breed' of jewellery/objects that combine handmade techniques with digital methods.
www.elizabeth.r.m.armour@gmail.com
www.elizabetharmour.com

Assa Ashuach

A product and furniture designer, Assa Ashuach is based in London. He is currently leading the cross-disciplinary MA Design Suite, London Metropolitan University and co-founder Digital Forming Ltd and UCODO ltd (User Co Design Objects). His OMI Light was one of the first 3D-printed products to be distributed to a mass market.
www.assaashuach.com
www.ucodo.com

Monika Auch (MA)

Monika's research is in visual artistic work, evident in her STITCH MY BRAIN project in Amsterdam. Research areas include, how digital technologies influence human creativity and if they shape new neuropathways; artistic research in defining 'the intelligence of the hand' via MRI scans and EEG registrations, mapping the sensitivity of fingertips and dexterity through scientific data.
www.monikaauch.nl

Farah Bandookwala (FB)

Farah uses a combination of process like 3D printing, haptic interfaces (Cloud9), and physical computing to create artefacts that depend on audience participation to reach their full potential. Although the objects she designs are reminiscent of exotic species in nature, they are in fact unique to CAD/CAM.
kiwifarah@gmail.com
www.farahb.com

Bathsheba

Bathsheba (Grossman) is a sculptor exploring the meeting of art, maths and bioforms. Her work is about how to exist in three dimensions: symmetry and balance, going from the origin to infinity, finding life and beauty in geometry. 3D printing is her main medium and she is very happy with it.
http://bathsheba.com

Alison Bell

Driven by curiosity, process and relative indifference to outcome, Alison needs to feel excited by an unknown before she makes a move. Digital has enabled her to do this, bringing with it unimagined delight, opening invisible doors and allowing new thoughts to flood in. This continues.
www.alisonbell.co.uk

Marloes ten Bhömer

Marloes ten Bhömer's work consistently challenges generic typologies of women's shoes through experiments with non-traditional technologies and material techniques. By reinventing the process by which footwear is made, the results serve as examples of new aesthetic and structural possibilities, while also to criticise the conventional status of shoes as cultural objects.
info@marloestenbhomer.com
www.marloestenbhomer.com

Blue Marmalade

Blue Marmalade, launched in 2002, sees itself as quintessentially British (just like marmalade) but with a difference (hence, blue marmalade). It produces fun, funky functional, classically contemporary, affordable, good and ecological designs that are bright, contemporary and eco-friendly. All off-cuts are either used for the smaller gift products or melted down and used again.
www.bluemarmalade.co.uk

Stephen Bottomley

Stephen trained at the Royal College of Art (1999–2001) and University of Brighton (MA 1996–98), including a key period on exchange at Rhode Island School of Design, USA (1998). For more than twenty years his creation of contemporary jewellery has reflected an emerging digital age with a quiet and confident minimalist beauty.
s.bottomley@ed.ac.uk
http://www.klimt02.net/jewellers/stephen-bottomley

Riccardo Bovo

The focus of Riccardo's design activity is the shift of paradigm between industrial revolution and informational revolution. Generative algorithms, interactive design systems, digital fabrication, these are the trigger concepts that drive his research and experimentation.
riccardo.bovo.design@gmail.com
www.riccardobovodesign.net

Dr Katie Bunnell

Katie Bunnell is Associate Professor of Design at Falmouth University and leads the Autonomatic, a group of practitioner-researchers exploring the relationships between digital technologies and craft practices. Katie is a ceramic designer-maker who disseminates her research nationally and internationally.
Katie.bunnell@falmouth.ac.uk
air.falmouth.ac.uk/research-groups/autonomatic

Gordon Burnett

Gordon's life in making is formed by collaborating with materials and tools, tempered by critical reflection with a desire to find 'rightness' to solutions and being inspired with the surprise and joy from working with others.
gordon@makingsenseresearch.net

Alison Counsell

Alison has found a way of incorporating her love of drawing into her metalwork via the photo-etching process. These stainless steel drawings are then manipulated into a wide range of 3D products. And sometimes she leaves the final making up to the recipient!
alisongcounsell@aol.com
www.alisoncounsell.com
www.wapentac.com

Lionel Dean (LD)

FutureFactories is a digital manufacturing concept for the mass individualisation of products achieved by combining geometry, data manipulation and CAD to create meta-designs with the capacity to change over time. His products are very much design led and about boundaries between art and design and between code and craft.
www.futurefactories.com

Charlotte De Syllas

Charlotte De Syllas is acknowledged as one of the finest artist-jewellers working in Britain today and has a portfolio of great originality and distinction. Her awards include the prestigious Jerwood Prize for Jewellery in 1995 and the Goldsmiths' Company Award in 2007.
info@charlottedesyllas.com
http://www.charlottedesyllas.com

Zachary Eastwood-Bloom (Z-EB)

A London-based artist with degrees from Edinburgh College of Art, Scotland and the Royal College of Art, London, Zac has exhibited with the Crafts Council, the Royal British Society of Sculptors, the 2012 Cultural Olympiad and at the V&A Museum and is a founder member of Studio Manifold.
z.eastwood-bloom@network.rca.ac.uk
www.zacharyeastwood-bloom.co.uk
www.studiomanifold.org

Michael Eden

Michael's work explores relationship between hand and digital tools enabling him to create work that can, to some small extent be used to address the future of craft. He is interested in how tacit knowledge and sensitivity to materials, processes and objects developed through extended practice affect and influence the approach to making using digital technology.
www.edenceramics.co.uk

Ben Gaskell

In 1990, Ben set up a specialist workshop in London making objects of the highest quality from hardstone, ornamental and precious stone and he won many Goldsmiths' Craft and Design Council Awards for his craftsmanship. Although interested in technology's possibilities for carving and shaping, he does not let technology bend him or seduce him from his purpose.
www.gaskellquartz.com

Beate Gegenwart

Questions relating to place, location and by extension dislocation and movement, are a continuous focus in Beate's work. Language occupies an important part in this inquiry, the idea of the 'translator' and the use of the 'mother tongue' as orientation, home and dwelling rather than physical location.
beate.gegenwart@btinternet.com

Otto Gunther

Otto gave up a career in medical IT in order to pursue creativity and passion in all its forms. His art draws heavily from the shapes he sees in the world around him and his fascination with geometric spaces. He currently resides in Atlanta.
ARTbyGUNTHER@gmail.com
http://www.etsy.com/shop/ARTbyGUNTHER

Kathryn Hinton (KH)

A jeweller and silversmith currently working in Edinburgh, Kathryn's work focuses on merging traditional skills with digital technology. At the Royal College of Art she began to research and explore the possibilities of computer-aided design and rapid prototyping, combining these processes with silversmithing techniques to create a new digital tool.
kath@kathrynhinton.com
www.kathrynhinton.com

Hot Pop Factory

Based in Toronto, Canada, Hot Pop Factory use emerging technologies to create intimate adornments for the body. Their collection, Stratigraphia, uses 3D printing to create bold and elegant jewellery that celebrates the stratified texture of additive manufacturing. The accretion of barely perceptible layers gives every piece a unique 'fingerprint' distinguishing it from others like it.
www.HotPopFactory.com

Tavs Jorgensen (TJ)

Danish-born Travs is a freelance designer and research fellow at the Autonomatic Research Group, University College Falmouth. His research is predominately focused on investigating how new computer interfaces can facilitate more personal and expressive aesthetics in artefacts created using digital tools.
http://www.tavsjorgensen.co.uk/

Markus Kayser

Markus is currently a research assistant inat the MIT Media Lab, Massachusetts, Cambridge, USA. He has joined the Mediated Matter Group, which focuses on 'how digital and fabrication technologies mediate between matter and environment to radically transform the design and construction of objects, buildings, and systems.'
m_kayser@mit.edu
contact@markuskayser.com
www.markuskayser.com

Birgit Laken

Following Ann Marie Shillito's development of haptic 3D-modelling software (Cloud9), Birgit, a jeweller and photographer, became interested in this new way of designing. She realised her work could advance further with 3D designing and 3D printing and that it was fun to find out about a whole new industrial process, while also make unique pieces too.
www.birgitlaken.nl

Lynne MacLachlan

Lynne is a jeweller and designer from Scotland living in London and a graduate of the Royal College of Art. Her PhD in the Design Transformations Group of the Open University is specifically concerned with the creative use of generative design in conjunction with digital fabrication.
www.lynnemaclachlan.co.uk
http://lynnemaclachlan.tumblr.com/

Geoff Mann

Geoff Mann is an artist, designer and lecturer. His studio questions and challenge what these practices mean today. From traditional craft to interactive digital media, Mann brings diverse skill-sets together to bring compelling narrative to life in engaging and innovative ways through working beyond the constraints of material or process.
enquiries@mrmann.co.uk
www.mrmann.co.uk

Dr Justin Marshall

Justin Marshall is Associate Professor of Digital Craft at University College Falmouth. He is a practising maker and researcher who has been investigating the role and significance of digital design and production technologies in craft and designer maker practice for over ten years.
Justin.marshall@falmouth.ac.uk

Gayle Matthias

Gayle Matthias is a member of Autonomatic research group at University College Falmouth where she is a Senior Lecturer on the BA Hons Contemporary Craft course. A practising glass artist, Gayle's work is part of the permanent collection of the V & A Museum.
gayle.matthias@falmouth.ac.uk
http://air.falmouth.ac.uk/research-groups/autonomatic

Alissia Melka-Teichroew

Designer, founder, creative director of byAMT Inc. where she develops her own designs and trieds to 'find the fine line between aesthetics and functional.' She trained at the Design Academy Eindhoven (NL), Rhode Island School of Design, USA, worked at IDEO and Puma International, and in 2010 she was visiting professor, Pratt Institute, New York.
wwww.byamt.com

Peter Musson

MA (RCA), Designer Silversmith, creating sophisticated silverware by smoothly combining the freedom of making by hand with controlled computer techniques. Peter embraces the unique decorative finish of both methods and displays evidence of this process, resulting in silver pieces that are as honest as they are elegant.
http://www.petermusson.com

Inge Panneels

Inge Panneels is an artist and academic whose current work uses mapping as a visual language and digital techniques (such as waterjet-cutting intricate fused glass pieces, and rapid prototyping models) to resolve complicated components, to achieve an aesthetic, otherwise impossible to achieve with traditional making techniques.
inge@idagos.co.uk
www.idagos.co.uk

David Poston

David's current work principally explores the use of computer-based making, welding and forging in combination with titanium, stainless steel, textiles and glass. His objective is to produce work that people enjoy the feel of wearing and that lightens the spirit of wearer, audience and, in the making, himself.
david.poston@iname.com
www.davidposton.net

Gilbert Riedelbauch (GR)

Gilbert is a silversmith, an academic and early user of digital technologies. He says of his work that as much as the mind links an idea with a design solution, it is the hand that connects design to the making process, forming the essence of craft. Skilful manipulation of tools and process – the basis of making – is equally important for digital work and requires an experienced hand. He is based in Canberra, Australia.
gilbert.riedelbauch@anu.edu.au
www.gilbertriedelbauch.com

Esteban Schunemann

Esteban Schunemann is an industrial designer and researcher at Brunel University. He is currently investigating the deposition of paste-like materials for RP in a number of design contexts, including medical, craft, product and open design. He lives in London.
esteban@schunemann.org

Jenny Smith

Jenny is a practising, award-winning artist in Edinburgh, exhibiting nationally and internationally with work in the Tate and National Galleries of Scotland. She has a degree in English Literature, a first-class degree in Drawing and Painting from Duncan of Jordanstone College of Art in Dundee and an MA in Multi-Disciplinary Print at UWE Bristol.
www.edinburghlaserstudio.org.uk

Sarah Silve

Sarah Silve, a lecturer on the Product Design programmes at Brunel University, is also course director BA Industrial Design and Technology. Her background is in silversmithing and jewellery and her research interests are in materials, rapid prototyping and laser material processing for creative outcomes. She is part of Brunel's Open Design and the Digital Economy research group.

SketchChair: (Greg Saul and Tiago Rorke)

SketchChair is a collaboration. Greg Saul is an industrial design graduate with a strong interest in using technology to enable innovative and new design. From his experience as a designer and researcher Tiago Rorke is usually immersed in prototyping and physical computing, tools and details. Tiago has since co-founded London-based Diatom Studio that is working on collaborative and open-source design tools based around digital fabrication.
www.gregsaul.co.uk
www.tiago.co.nz
www.sketchchair.cc

Johanna Spath (JS) and Johannes Tsopanides

Johanna and Johannes, who are based in Berlin, are SHAPES iN PLAY. Playful and explorative use of generative product design enables them to create new solutions for concepts and shapes.
mail@shapesinplay.com
www.shapesinplay.com

Anthony Tammaro

Anthony is a new media artist operating at the Art/Design/Craft intersection. His most recognisable work leverages 3D software expertise with 3D-printing processes. He creates novel solutions to design problems related to the body as site. Presently researching a new approach that incorporates technology into craft-based studio practice to culminate in a new exciting body of work.
www.anthonytammaro.com
http://clutchworks.blogspot.co.uk

Cathy Treadaway (CT)

Dr Cathy Treadaway is an artist, educator and writer. She is Reader in Creative Practice at Cardiff School of Art and Design, Cardiff Metropolitan University and works with artists, designers and industry to investigate the development and use of digital technologies to support creativity.
www.cathytreadaway.com

Christopher Tipping

Christopher explores themes of place making in the public realm. Commissioned projects are exclusively site- and context-specific in origin and are shaped and underpinned by extensive research and consultative practice.
www.axisweb.org/artist/christophertipping

Unfold

Defined as 'to be revealed gradually to the understanding', Unfold was founded in 2002 by Claire Warnier and Dries Verbruggen. The Antwerp-based duo developed a strong multidisciplinary background in design, technology and art and often collaborates with a vast network of kindred spirits and specialists.
www.unfold.be/pages/projects

Daniel Widrig (DW)

Architect, product and furniture designer, sculptor and artist. After working with Zaha Hadid Architects for several years, he went on to establish his own studio where he continues to work on experimental concepts not only in architecture but also in the world of fashion with his collaboration with fashion designer, Iris Van Herpen.
www.danielwidrig.com

Jae-won Yoon

Jae-won was born in Seoul, South Korea. In 2009 he completed his MA in Jewellery and Silversmithing at Edinburgh College of Art. He now now holds an academic position at Gwangju University in South Korea.
jw0409@gwangju.ac.kr

Initials of authors supplied where the author contributed to the 'Author questionnaire' referred to throughout the book, see page 11.

Glossary

3D PRINTING: computer-controlled process of building up a 3D form by adding material. Other terms include layer manufacture, additive manufacture and rapid prototyping. 3D computer graphics (CG): different from CAD, 3D modelling and solid modelling. Many CG graphics engines use 3D vector triangles to represent the surface of an object, as these are very effective for simulating 3D in real time on 2D pixel-based screens. Full-colour, surface-textured images are continually re-rendered onto very fast changing entities using another part of the graphics engine. As these vector/pixel entities are not encumbered with full 3D geometry they are not whole three-dimensional entities which can be 3D printed and milled. Further complex processing programs are required to convert them into solid models capable of being 'sliced' into layers for 4/5/6 axis milling and 3D printing.

ADDITIVE MANUFACTURE: See 3D printing

ALGORITHM: A process, set of rules or instructions to be followed step by step and used in programming software to perform certain tasks, whether in calculations, for solving a problem, or completing an automated task. (http://www.techterms.com/definition/algorithm)

AMF (ADDITIVE MANUFACTURING FORMAT OR ADDITIVE MANUFACTURING FILE): more accurately defines a 3D virtual object using a simpler, more efficient file structure than currently used in the industry to process a form for machining or 3D printing. AMF will eventually replace STL file format. (see STL)

BOOLEAN: used here as a 3D modelling function where volumetric objects are either subtracted one from another, unified into one surface or both subtracted, leaving the part where they overlapped as a new object form.

CAD (COMPUTER-AIDED DESIGN): engineering-based software for designing in two and in three dimensions.

CAD/CAM: computer-aided design/ numerical controlled machining.

CAM (COMPUTER-AIDED MACHINING OR COMPUTER-AIDED MANUFACTURING): a type of computer application that helps automate the manufacturing processes.

CLOUD COMPUTING: providing computing resources (hardware and software) as a service over a network, typically the Internet. 'Cloud' refers to the cloud-shaped symbol used to represent the complex infrastructure in system diagrams. Users' data is stored on servers at a remote location where they can also access cloud-based applications and computation. Types of public cloud computing include 'Software as a service'. Examples of cloud computing are Apple's iCloud and Dropbox. (http://en.wikipedia.org/wiki/Computer_cloud)

CNC (COMPUTER NUMERICAL CONTROL): (http://www.cnccookbook.com/CCDictionary.htm)

CROWDFUNDING: campaigns on platforms such as Kickstarter, IndieGoGo, Pozible where individuals and small businesses use the Internet to raise funds and investments from family, friends, the community and the public – in fact, anyone interested in their venture.

DESKTOP 3D PRINTER: lightweight machines providing small businesses with the capability to do their own concept prototyping and building models in-house and with reasonable standards of quality.

DIGITISING PEN: a hand-held stylus, sometimes used with a tablet, whose point is enabled to provide accurate positioning data that is graphically translated on the monitor, whether as writing, drawing or for selecting functions.

DXF (DRAWING EXCHANGE FORMAT): commonly supported in CAD programs although it can be inefficient and difficult.

FLOW: a deeply qualitative experience. Mihaly Csikszentmihalyi's 'flow' word is very apt for describing experience of being totally immersed in doing something, whether making, rock-climbing, playing a piano, etc., of being carried by a current, spontaneously and effortlessly, forgetting time and having no fear of being out of control. The situation is exhilarating, balanced on that edge between knowing your capabilities yet having a tinge of anxiety about your capacity to complete the task successfully.

FORCE FEEDBACK: by using mechanical devices that provide a physical sensation of resistance (force feedback) to the user, physical attributes such as weight and stiffness can be applied to virtual objects allowing the user to interact directly using touch. (see haptics)

G-CODE: the main programming language used to control CNC machines. Other examples are Heidenhain and Mazak.

GENERATIVE DESIGN: a design method using a computer program whereby the product (image, sound, pattern, model, animation etc.) is generated by a set of rules, an algorithm, to create variations swiftly, either from a structured scheme or the use of random numbers, whereby designs develop as genetic variations through mutation and crossovers. (http://en.wikipedia.org/wiki/Generative_Design)

GOLDSMITHS' CRAFT AND DESIGN COUNCIL: a UK industry body that promotes design and craftsmanship within the UK jewellery industry and allied crafts. The Council runs the annual Craftsmanship & Design Awards open to anyone based in the UK and involved in silversmithing, jewellery and the allied crafts. (http://www.craftanddesigncouncil.org.uk/)

GRASSHOPPER: graphical algorithm editor that is also tightly integrated with Rhino's 3D modelling tools. (http://www.grasshopper3d.com)

GUI (GRAPHICAL USER INTERFACE): GUI has replaced typing commands. It uses a visual cursor (controlled by a mouse or stylus) to move about the screen, clicking on standard graphical icons, slider bars etc, to select functions.

HAPTICS: tactile feedback technology that takes advantage of the sense of touch by applying forces, vibrations, or motions to the user, giving the sensation of 'virtual touch' experienced through the force feedback device. It has been described as 'doing for the sense of touch what computer graphics does for vision'.

IGES (INITIAL GRAPHICS EXCHANGE SPECIFICATION): one of the most commonly supported formats in the computer-aided designing and manufacture industry.

KERF: the width of a groove made by a cutting tool.

LASER (LIGHT AMPLIFICATION BY STIMULATED EMISSION OF RADIATION): is used in industry for precise cutting by melting, burning or vaporising a wide range of materials. The main types of lasers are:
Carbon dioxide (CO_2) lasers: low powered, cuts paper to fabrics to acrylics (up to 20 mm), wood (plywood, wood and MDF: 25 mm thickness), etc. High powered, cuts thin sheet metals, mild steel (up to 20 mm) and for engraving metal and hard materials.
High-pressure nitrogen: cuts stainless steel (15 mm thickness): bright, clean cut, oxide-free edge, ideal for most applications, including welding, no further finishing required.
Nd:YAG (Neodymium-doped yttrium aluminum garnet): critical tolerance work, especially cutting thin foils, fine details, hard and exotic metals (tool steels up to 12.5 mm, aluminium up to 10 mm, titanium, niobium).

LAYER MANUFACTURE: another term for, and type of, 3D printing.

MAKERBOT: name of a personal 3D printer based on original open-source information, and the name of the US company manufacturing and selling the printer.

PERSONAL 3D PRINTER: spin-offs from the original RepRap open source 3D printer kit, developed by Adrian Bowyer at Bath University, UK. Examples: MakerBot, Maxit, Ultimater and there are many more low-cost 3D printers either sold in kits or pre-built.

PONOKO: Ponoko was launched in 2007 at Techcrunch40 as the world's first digital making system and the world's easiest making system, offering 2D and 3D manufacture, great web resources, shop, basic software for designing and affiliated companies such as RazorLab in London, Formulator in Germany.

'PROCESSING' SOFTWARE: founded and developed by Ben Fry and Casey Reas in 2001 as 'software sketchbooks' for learning the fundamentals of computer programming within a visual context. It is so straightforward that anyone who wants to can create images, animations and interactions. It has evolved into a tool that thousands of people are learning and using for playing, exploring, prototyping, and for generating finished professional work. Its sheer usability enables designers to work visually, exploring and experimenting in a risk-free environment.

PROPRIOCEPTIVE SENSE: the unconscious perception of movement and spatial orientation arising from stimuli within the body, detected by nerves as well as by the semicircular canals of the inner ear that enable us to interact three dimensionally with assurance.

RP (RAPID PROTOTYPING): Lemmings 1997, 'A process that automatically creates a physical prototype from a 3D CAD-model in a short period of time.' A family of technologies that produce physical three-dimensional forms from computer-generated digital models. The ones mentioned here are layer manufacture, **CNC** manufacturing systems (computer numerical controlled machining), and **reverse engineering (Scanning)**.

REPRAP: the original open-source information for constructing a personal 3D printer, invented by Adrian Bowyer (see Personal 3D printer and MakerBot, and page 157).

SINTERING: the careful, controlled heating of powder or granules (metallic powder, for example) to form a coherent mass by melting.

STL (.STL): the most commonly used and preferred format for CNC machining where surfaces are represented by triangulation and the resolution can be defined. STL is specifically designed to capture precisely, as a long list of triangles, the 3D model's exact shape and form, normalising, regulating, and ordering the surfaces of triangles so that the model can be processed smoothly and sliced for 3D printing and machining. The STL file format will eventually be replaced, possibly by **AMF (additive manufacturing format or additive manufacturing file)**, which more accurately defines objects using a simpler and more efficient file structure.

SUBTRACTIVE MANUFACTURING: removal of material from a solid block by mechanical means such as milling and lathing.

THREE-DIMENSIONAL PRINTING: uses inkjet heads to deposit glue to locally bind layers of powder.

WIMP INTERFACE (WINDOWS, ICON, MOUSE, POINTER): the standard means with which we interact digitally on computer.

Bibliography and further reading

Adamek, Anna, Assistant to the Curator at the Canada Science and Technology Museum Decal Collection and information, also transferring techniques to apply decals: water release, cement mounting, or a pressure sensitive process, used mostly for porcelain decoration, Canada Science and Technology Museum. Available at: http://sciencetech.technomuses.ca/english/collection/industrial6.cfm.

Lumps of Geometry by Glen Adamson (2nd August 2010). Available at: www.vam.ac.uk/blogs/sketch-product/lumps-geometry.

Anderson, Chris, *Makers* (London: Random House Business Books, 2012).

Ashby, M. and Johnson, K., *Materials and Design: the Art and Science of Material Selection in Product Design* (Oxford: Butterworth-Heinemann, 2009).

Baker, Robin, *Designing the Future: The Computer transformation of Reality* (London: Thames & Hudson, 1993).

Words by Fred – Vision and Reality by Fred Baier. Available at: http://www.fredbaier.com/words/by-fred/p/vision-reality (accessed 12 December 2011).

Balter, Michael, 'Tool Use is Just a Trick of the Mind', *ScienceNOW*, 2008.

Barthes, Roland. 'Cy Twombly: Works on Paper' and 'The Wisdom of Art' in *The Responsibility of Forms* (Berkeley: University of California Press, 1985).

Beylerian, George M., Osborne, Jeffrey J., Kaufman, Elliott, Steelcase Design Partnership, Cooper-Hewitt Museum (New York). *Mondo Materials: materials and ideas for the Future* (New York: Overlook Press, 2001).

Bottomley, S., Goodwin, D. Paper, 'Something Old/Something New: The Marriage of Digital – Craft,' Challenging Craft International Conference, Aberdeen, 2004. Available at: http://www2.rgu.ac.uk/challengingcraft/ChallengingCraft/pdfs/bottomleygoodwin.pdf.

Bottomley, S., 're-worthit!' authored catalogue, 2009. Peer reviewed by Dr David Humphries, exhibition of work by 22 artists marking centenary of Woolworths and its sudden closure in 2008. Printed versions of the catalogue are for sale at www.blurb.com/bookstore.

Bunnell, K. and Marshall, J. 'Developments in post-industrial manufacturing systems and the implications for craft and sustainability.' Paper presented at Making Futures Conference, Plymouth College of Art, 2009.

Campbell, J. R., New Craft Future Voices Conference, Dundee, Scotland, 2007.

Cooper, Alan, *Inmates Are Running the Asylum: Why High-Tech Products Drive Us Crazy and How to Restore the Sanity* (US: Sams Publishing, 2004).

Dormer, P., 'Craft and the Turing Test for Practical Thinking' in Dormer, P. (ed.) *The Culture of Craft* (Manchester: Manchester University Press, 1997).

Debold, Elizabeth, 'An interview with Dr Mihaly Csikszentmihalyi', *EnlightenNext Magazine*, Spring-Summer 2002. Available at: http://www.enlightennext.org/magazine/j21/csiksz.asp?page=3.

'The Printed World' (Febuary 2011) *Economist Online*. Available at: http://www.economist.com/node/18114221.

Francisco, Scott. Response to 'The Printed World' (Febuary 2011) *Economist Online*. Available at: https://www.economist.com/user/4431648/comments.

Edwards, Clive, *The Castration of Skill. Obscure Objects of Desire* (London: Crafts Council, 1997).

Flahery, Joseph. 'Personal Fabrication for Dummies' (19 October 2008). Available at: http://replicatorinc.com/blog/2008/10/personal-fabrication-for-dummies.

'The Innovation Paradox: How Innovation Products Threaten the Innovation Process' by Scott Francisco (November 2010), *Reconstruction: Studies in Contemporary Culture*, Vol. 10, No. 2, 2010. Available at: http://reconstruction.eserver.org/102/recon_102_francisco01.shtml.

'Gartner's 2012 Hype Cycle for Emerging Technologies Identifies "Tipping Point" Technologies That Will Unlock Long-Awaited Technology Scenarios', press release from Gartner (16 August 2012). Available at: http://www.gartner.com/it/page.jsp?id=2124315.

'Momentum'. Exhibition catalogue: curated by Beate Gegenwart in conjunction with Charlotte Kingston at Craft in the Bay, Cardiff, 2011.

Gell, A., 'The Technology of Enchantment and the Enchantment of Technology' in Candlin, F. and Guins, R. (eds.) *The Object Reader*. (London/New York: Routledge, 2009, pp. 208–28).

Brainwashed: seven ways to reinvent yourself, by Seth Godin. Available at: http://www.sethgodin.com/sg/docs/brainwash.pdf.

Hanks, K., and Belliston, L. *Rapid Viz, a new method for the Rapid Visualization of Ideas* (Canada: Crisp Publications, 1990).

Harris, Dr Jane."Crafting' 'Computer Graphics': A Convergence of 'Traditional' and 'New' Media", *The Journal of Cloth and Culture*, 3 (1), 2005, pp. 20–34. Abstract available at: http://core.kmi.open.ac.uk/display/8769869.

Heidegger, M. *Basic Writings*, (D. Farrell Krell (ed.), San Francisco: Harper and Row. 1977)

Hopkinson, N, Hague, R. J. M., Dickens, P.M, *Rapid Manufacturing – An Industrial Revolution for the Digital Age* (Chichester: John Wiley & Sons, 2006).

Kettley, Sarah. 'Distribution - craft and speckled computing', Craft Australia Research Centre: WearNow Symposium , Canberra in February, 2007. Available at: http://www.sarahkettleydesign.co.uk/sarahkettley/publications_sarah_kettley__files/craftAustralia.pdf.

Kettley, S. and Smyth, M. (2006) 'Plotting affect and premises for use in aesthetic interaction design: towards evaluation for the everyday'. Proceedings of HCI UK, Vol.1. (Human Computer Interaction). London September 11–15.

Smith, M. K. 'David A. Kolb on experiential learning', *Encyclopedia of Informal Education*, (2001). Available at: http://infed.org/mobi/david-a-kolb-on-experiential-learning/.

Kolb, D. A., *Experiential Learning: Experience as the Source of Learning and Development*. (Englewood Cliffs, NJ: Prentice Hall, 1984). Full statement and discussion of Kolb's ideas concerning experiential learning. Chapters deal with the foundation of contemporary approaches to experiential learning; the process of experiential learning; structural foundations of the learning process; individuality in learning and the concept of learning styles; the structure of knowledge; the experiential learning theory of development; learning and development in higher education; lifelong learning and integrative development.

Lambert, I. and Firth, R., 'Pencils Don't Crash', Engineering and Product Design Education Conference, Salzburg, 2006.

Lefteri, C., *Making it: Manufacturing techniques for product design* (London: Laurence King, 2007). Deals predominantly with mass-production techniques, some well established some new.

Lewton-Brain, Charles. 'Some thoughts on

computer use in the Metals/Jewelry Field', 1990–1996.

Margetts, Martina, 'Action not Words', In Charney, Daniel, *Power of Making* (London: V&A Publishing and the Crafts Council, 2011, p. 43.)

Marshall, J., Unver, E. and Atkinson, P., 'AutoMAKE Generative systems, digital manufacture and craft production' Paper presented at Generative Art Conference, Milan, 11–14 December, 2007.

Marshall, Justin (2002) 'Craft in the twenty-first century: theorising change and practice' (Craft and Technology Strand). Craft and Technology Conference, Edinburgh College of Art.

Marshall J., Project 1, 'Computers and Creativity (investigating just how useful computer-aided design can be for studio potters.)' The research project was funded by the AHRB and carried out at Bath Spa University College under the supervision of Felicity Aylieff. SUMMER 2001

Marshall J. Project 2 'The employment of CNC milling in the Production of Low relief tiles', Article for Ceramic Technical. 19/2/02. This work was produced as part of a collaborative project with Dr Katie Bunnell, Reader in Design Research at Falmouth College of Arts and was funded by the Arts and Humanities Research Board. The author would also like to thank Tavs Jorgensen (mould making) and Dartington Pottery (pressing and firing) for their help and support.

McCullough, M., *Abstracting Craft: the Practiced Digital Hand.* (Cambridge, MA: MIT Press, 1996).

McKeough, Tim, 'How Multimillionaire Zita Cobb Plans to Turn a Tiny Canadian Island Into an Arts Mecca', *Fast Company* (15 December 2012). Available at: www.fastcompany.com/1702241/how-multimillionaire-zita-cobb-plans-turn-tiny-canadian-island-arts-mecca (accessed 28 November 2011).

Polyani, M. *Personal Knowledge* (London: Routledge & Kegan Paul, 1958).

Polanyi, Michael, *The Tacit Dimension* (New York: Doubleday Anchor, 1967).

Postman, Neil, *Technology: The surrender of culture to technology* (New York: Vintage Books, 1993).

Press, M. 'What has craft given us?' *Crafts* no. 227, November/December 2010, pp. 104–6. See also Mike Press website http://mikepress.wordpress.com/ 06.12.10.

Pye, David, *The Nature and Art of Workmanship* (Cambridge: Cambridge University Press, 1968, pp. 17–29).

Ravetz, A. and Webb, J., 'Migratory Practices: introduction to an impossible place? *craft + design enquiry,* vol. 1 (2009). Available at: www.craftaustralia.org.au/cde.

Ravetz, Amanda and Webb, Jane, 'The Craft of Enlightenment', Manchester Craft Rally, Manchester Metropolitan University, 8 December, 2010.

Smith, M. K., 'Donald Schön: learning, reflection and change', (2001, 2011) *the encyclopedia of informal education*, Available at: www.infed.org/thinkers/et-schon.htm (accessed 2nd January 2012).

Schön, D, *The Reflective Practitioner. How professionals think in action* (London: Temple Smith, 1983). Influential book that examines professional knowledge, professional contexts and reflection-in-action. Examines the move from technical rationality to reflection-in-action and examines the process involved in various instances of professional judgement.

Schmitt, R.F., Thews, G., 'Physiologie des Menschen,' *Springer Verlag.* Available at: http://www.zoologie-skript.de/gruvo03/zns/scortex.htm.

Schwarz, Mary and Yair, Dr Karen. 'Making VALUE: craft and the economic and social contribution of makers', June 2010. Available at: http://www.craftscouncil.org.uk/files/file/7cec2fd1e3bdbe39/making_value_full_report.pdf.

Sennett, R., *The Craftsman* (London: Penguin Books, 2009).

Smith, J., 'Digital Calligraphy', Practitioners Illustrated Presentation at Cutting Edge Lasers and Creativity Symposium, Loughborough University, 2009.

Wolfe, Josh. See: http://www.forbes.com/sites/joshwolfe/2012/08/08/3d-printing-with-peter-weijmarshausen/.

http://thebodyandtechnology.wordpress.com and http://thebodyandtechnology.wordpress.com/further-reading/. An excellent blog exploring an embodied approach to synthesising traditional craft and design processes with new and emerging technologies.

General links and websites

This is a totally eclectic list collated in the process of researching and searching. I hope it is of interest and also useful as jumping-in places!

2D Plotter Cutters:

- http://en.wikipedia.org/wiki/Plotter
- http://blog.ponoko.com/2010/07/26/make-prototypes-for-laser-cutting-at-home/

Personal 3Dprinting, RepRap

- http://reprap.org/wiki/Main_Page
- http://adrianbowyer.net/1_3_Publications.html

3DPrinting service providers:

- www.Shapeways.com
- http://i.materialise.com
- www.Sculpteo.com
- http://www.100kgarages.com
- Mcor's paper 3D printers: http://www.staples.nl/
- www.ponoko.com
- www.3dprint-uk.co.uk

Interesting 3Dprint bits and bobs:

- http://www.youtube.com/watch?v=-yv-IWdSdns&feature=player_embedded
- http://en.wikipedia.org/wiki/3D_printing
- UK distributors for equipment to 3D printing in gold and titanium: www.estechnology.co.uk
- http://www.3dprint-uk.co.uk/3d_printing_costs.html
- http://fabbaloo.com/blog/2011/12/21/the-list-of-personal-3d-printers-2011.html#.UCoAdp1lSSo
- 3D printing cream cheese and insects: http://sciencegallery.com/edible/technology/insects-au-gratin,
- http://info.uwe.ac.uk/news/uwenews/news.aspx?id=2211
- http://www.mcortechnologies.com/leather-like-sculptures-created-with-paper-3d-printing/
- 3D printing 'wood flour': http://open3dp.me.washington.edu/2011/04/woodnt-you-know-it-3dp-in-wood/
- www.ponoko.com/make-and sell/design-it-yourself
- http://blog.ponoko.com/2012/03/14/10-excellent-tips-tutorials-on-3d-printing/
- http://support.ponoko.com/entries/20966162-rich-s-top-ten-list-of-stuff-ups-in-3d-printing
- http://www.ponoko.com/make-and-sell/show-material/357-3d-printed-glazed-ceramic#main-image

3D print post-processing:

- http://www.3dprintcraft.wikispot.org/

3D print Bloggers:

- Joris Peels
- Fabbaloo
- 3ders

3D scanning:

- http://www.rapido3d.co.uk/
- http://www.inition.co.uk
- http://www.zcorp.com/Products/3D-Scanners/spage.aspx,
- https://www.nextengine.com/indexSecure.htm
- http://www.escan3d.com/?gclid=CMbSoojps5YCFQNHFQodFUhiLQ

Low-cost scanning:

www.david-laserscanner.com/
www.123dapp.com:

CNC milling:

- http://straightlinedesigns.wordpress.com/2011/10/13/the-bubble-cabinet-apple-cabinet/
- eMachine Shop: http://www.emachineshop.com/,
- Tech Shop: http://www.techshop.ws
- Craftsman Compucarve: http://www.sears.com/shc/s/p_10153_12605_00921754000P
- http://studiomill.co.uk/software/
- http://www.rolanddg.com/news/nr111005_im-01.html
- http://www.delcam.com/delcamhome.htm

CNC embroidery:

- Singer: http://www.singerco.com/futura,
- Brother: http://www.brother-usa.com/HomeSewing/ProductList.aspx?cat=embroidery&W
- T.svl=EmbroideryTopNav
- Toyota: http://www.pantograms.com/machines.asp?p=m1

Downloadable models:

- http://grabcad.com/dizingof
- http://thingiverse.com/dizingof
- http://n-e-r-v-o-u-s.com/shop/line.php?code=5

More interesting bits:

- InsideOut Exhibition: http://shura.shu.ac.uk/4392/4/insideout_media_release.pdf
- http://formero.com.au/photo-gallery/photos/12
- http://www.core77.com/blog/object_culture/the_melonia_shoe_a_worlds_first_wearable_3d_printed_footwear_15995.asp
- http://surveys.peerproduction.net/2012/05/manufacturing-in-motion/6/
- http://www.freedomofcreation.com/for/freedom-of-creation-enters-customization-with-kickstarter
- http://www.designboom.com/weblog/cat/8/view/20710/lasercut-nori-for-designer-sushi.html
- Excellent blog about craft and applied art: http://www.craftresearch.blogspot.com/
- Online crowd-funding platform: www.kickstarter.com.
- JST ERATO Igarashi Design UI Project: www.designinterface.jp/en/overview/

Electronic stuff:

- Arduino: http://www.adafruit.com/tutorials
- Mztek (@MzTEK): (Women artists, tinkerers and techies! London): http://www.mztek.org

FabLabs, Hacklabs and workspaces:

- http://fablab-international.org/
- http://wiki.fablab.is/wiki/Portal:Labs
- Denmark: http://fablabdanmark.dk/
- Glasgow: www.maklab.co.uk
- Lisbon, Portugal: http://fablabedp.edp.pt/ -
- London: www.mak3d.com/

HUBS:

- MetropolitanWorks
- The European Ceramic Work Centre (.ekwc), 's-Hertogenbosch, the Netherlands,: http://www.ekwc.nl.

Learning, courses:

- Anarkik3D: www.anarkik3d.co.uk
- Rhino www.simplyrhino.co.uk
- Jenny Smith: http://www.edinburghlaserstudio.org.uk
- Video library: lynda.com

Laser cutting:

- http://thelasercutter.blogspot.co.uk/
- Light-weight laser cutting: http://www.lasercutit.co.uk/
- http://www.cutlasercut.com/
- Silver laser cutting: www.capital-lasers.co.uk
- http://flickrhivemind.net/Tags/lasercutting/Interesting
- Ponoko http://www.ponoko.com

Materials:

- http://www.ahrc.ac.uk/News-and-Events/Watch-and-Listen/Pages/3D-Printing-in-Ceramics.aspx
- Mondo Materials: materials and ideas for the Future
- Coloured fiberboard material, produced in Portugal by Investwood: www.valchromat.pt/
- http://i.materialise.com/blog/entry/3d-printing-in-titanium-still-going-strong
- 3D printed ceramics: http://figulo.com/figulo/Home.html
- http://3dceram.com/en/category/luxe/applications-luxe/
- http://openmaterials.org/2010/02/19/open3dps-recipes/

Pixels:

- http://www.cambridgeincolour.com/tutorials/digital-camera-pixel.htm

Photo etching:

- Silver photo etching: http://www.chempix.com/
- www.photofab.co.uk

QR code generator:

- http://qrcode.kaywa.com/

Sheet metal forming:

www.ifm.cam.ac.uk/sustainability/projecs/incremental

Pricing:

http://www.3dprint-uk.co.uk/3d_printing_costs.html
http://www.3DprintingPriceCheck.com

Waterjet cutting:

www.tmcwaterjet.co.uk
www.waterjets.org
http://www.flowwaterjet.com

Textile printing:

http://www.catdigital.co.uk/

Software lists:

- http://www.3dprinter.net/directory/3d-modeling-software
- http://fabbaloo.com/3d-resources#commsoftware
- File Types and extensions: http://www.fileinfo.com/filetypes/common
- Scripting software for designing: http://www.generatorx.no/
- Scripting/coding: Java-based programming language 'Processing' (www.processing.org)
- Inspirational and scary: http://ds13.uforg.net/ using scripting for architecturally complex forms

Index

Note: pages with relevant illustrations are listed in italics